RADIOGRAPHIC TESTING CLASSROOM TRAINING BOOK

Written for ASNT by

Jean Staton
The Ocean Corporation

The American Society for Nondestructive Testing, Inc.

Published by The American Society for Nondestructive Testing, Inc.
1711 Arlingate Lane
Columbus, OH 43228-0518

ASNT exists to create a safer world by promoting the profession and technologies of nondestructive testing.

ISBN 1-57117-120-7

Printed in the United States of America

Library of Congress Cataloging-in-Publication Data
Staton, Jean.
  The ASNT personnel training publications radiographic testing classroom training book / written for ASNT by Jean Staton.
      p. cm.
  Includes bibliographical references and index.
  ISBN 1-57117-120-7
  1. Radiography, Industrial. 2. Nondestructive testing. I. American Society for Nondestructive Testing. II. Title.

  TA417.25.S73 2005
  620.1'1272--dc22
                                    2005022044

First printing 09/05
Second printing with revisions 02/08
Third printing with revisions 10/09
Fourth printing with revisions 08/11

Acknowledgments

A special thank you goes to the following technical editors who helped with this publication:
David Culbertson, El Paso Corporation
Ron Kruzic, Chicago Bridge & Iron Company
Michael V. McGloin, Hellier

A special thank you goes to the following technical reviewers who helped with this publication:
Brian MacCracken, Pratt and Whitney
Robert Plumstead, Testwell Laboratories, Inc.
William Plumstead, Sr., Plumstead Quality and Training Services
David Quattlebaum, Quattlebaum Consultants
George Wheeler, Materials & Processes Consultants

A special thank you goes to the following corporate reviewers who helped with this publication:
Jerry Fulin, El Paso Corporation
Jim Houf, The American Society for Nondestructive Testing
Don Locke, Karta Technologies, Inc.
Charlie Longo, The American Society for Nondestructive Testing
S.O. McMillan, BAV Quality Assurance

A special thank you goes to the following SI reviewer who helped with this publication:
Frank Iddings

The Publications Review Committee includes:
Chair, Sharon I. Vukelich, University of Dayton Research Institute
B. Boro Djordjevic, Johns Hopkins University
Mark A. Randig, Cooperheat-MQS, Inc.

Ann E. Spence
Educational Materials Editor

Foreword

The American Society for Nondestructive Testing, Inc. (ASNT) has prepared this series of *Personnel Training Publications* to present the major areas in each nondestructive testing method. Each classroom training book in the series is organized to follow the Recommended Training Course Outlines found in *Recommended Practice No. SNT-TC-1A*. The Level I and Level II candidates should use this classroom training book as a preparation tool for nondestructive testing certification. A Level I or Level II may be expected to know additional information based on industry or employer requirements.

Table of Contents

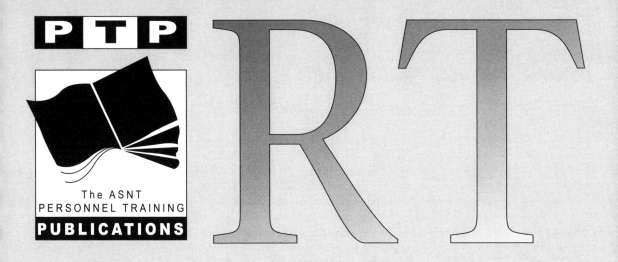

PTP

The ASNT PERSONNEL TRAINING PUBLICATIONS

RT

LEVEL I

Chapter 1

Introduction to Radiographic Testing

NONDESTRUCTIVE TESTING PERSONNEL

One of the most important aspects of any nondestructive testing method is that all personnel be trained, qualified and certified. Personnel must be familiar with the technique, equipment, test objects and how to interpret the results.

RADIOGRAPHY

Radiography is used to test a variety of products, such as castings, forgings and weldments. It is also used heavily in the aerospace industry for the detection of cracks in airframe structures, detection of water in honeycomb structures and for foreign object detection. Test objects are exposed to X-rays or gamma rays, and an image is processed on film or displayed digitally. Radiographic testing personnel set up, expose, process the film or digitally process the signals, and interpret the images in accordance with written codes or specifications.

Advantages of Radiographic Testing
The advantages of radiographic testing include the following.

1. Radiography can be used with most materials.
2. Radiography can be used to provide a permanent visual record of the test object on film or a digital record for subsequent display on a computer monitor.
3. Radiography can reveal some discontinuities within a material.
4. Radiography discloses fabrication errors and often indicates the need for corrective action.

Limitations of Radiographic Testing
Radiography's limitations include physical and economical considerations.

1. Radiation safety procedures must always be followed.
2. Accessibility can be limited. The radiographer must have access to both sides of the test object.
3. Discontinuities that are not parallel with the radiation beam are difficult to locate.
4. Radiography is an expensive testing method.

5. Film radiography is a time consuming testing method. After taking the radiograph, the film must be processed, dried and interpreted.
6. Some surface discontinuities may be difficult, if not impossible, to detect.

TEST OBJECTIVE

The objective of radiographic testing is to ensure product reliability. This can be accomplished based on the following factors.

1. Radiography enables the technician to view the internal quality of the test object or show the internal configuration of the components.
2. Radiography reveals the nature of the test object without impairing usefulness of the material.
3. Radiography reveals structural discontinuities, mechanical failures and assembly errors.

Performing the actual radiographic test is only part of the procedure. The results of the test must be interpreted to acceptance standards, and then the test objects are accepted or rejected.

Safety Considerations

Radiographic testing processes require X-ray and gamma ray sources that generate great amounts of radiation. Radiation can cause damage to the cells of living tissue, so it is essential that personnel be adequately aware and protected. Radiographic testing and quality assurance personnel must be continually aware of the radiation hazard and mindful of safety regulations.

Specially designed meters have the ability to detect X-radiation and gamma radiation. Radiation meters, called *survey meters*, are crucial instruments because radiation cannot be detected by sight, sound, touch, smell or taste. Strict observance of state and federal safety regulations is mandatory. Many jurisdictions require separate radiation safety certification to ensure technicians are cognizant of the safety regulations.

QUALIFICATION

It is imperative that personnel responsible for radiographic testing are trained and qualified with a technical understanding of the test equipment and materials, the test object and the test procedures.

The American Society for Nondestructive Testing (ASNT) has published guidelines for training and qualifying nondestructive testing personnel since 1966. These are known as *Personnel Qualification and Certification in Nondestructive Testing:*

Recommended Practice No. SNT-TC-1A. The *Recommended Practice No. SNT-TC-1A* describes the knowledge and capabilities of nondestructive testing personnel in terms of certification levels.

ASNT CP-189 was approved by the ASNT Board of Directors in 1989 as a standard for the qualification and certification of nondestructive testing personnel. The intent was to produce a new document that provided strict requirements rather than simply guidelines. ASNT obtained ANSI accreditation to process this document through a consensus balloting process that would recognize *ASNT CP-189* as a national standard. The first successful consensus document became *ANSI/ASNT CP-189-1991.*

ANSI/ASNT CP-189-2001 is similar to *SNT-TC-1A* in terms of training, experience and examinations, however, the standard provides minimum requirements that do not permit changes. Several significant differences were introduced to strengthen the NDT personnel qualification and certification program, which include the following.

1. Employer certification requirements and ASNT NDT Level III certification in the method.
2. Instructor for training must meet qualifications of the standard.
    a. ASNT Level III certificate.
    b. Bachelor of Science in engineering, physical science or technology with knowledge of nondestructive testing method.
    c. NDT Level II with at least ten years of experience.

Levels of Qualification

There are three basic levels of qualification applied to nondestructive testing personnel and used by companies that follow *Recommended Practice No. SNT-TC-1A* and *ASNT CP-189*: Level I, Level II and Level III.

An individual in the process of becoming qualified or certified to Level I radiographic testing is considered a trainee. A trainee does not independently conduct tests, interpret, evaluate or report test results of any nondestructive testing method. A trainee works under the direct guidance of certified individuals.

Qualification for Level I

Level I personnel are qualified to perform the following tasks.

1. Perform specific calibrations and nondestructive tests in accordance with specific written instructions.
2. Record test results. Normally, the Level I does not have the authority to sign off on the acceptance and completion of the nondestructive test unless specifically trained to do so with clearly written instructions.
3. Perform nondestructive testing job activities in accordance with written instructions or direct supervision from Level II or Level III personnel.

Qualification for Level II

Level II personnel are qualified to perform the following tasks. A Level II must be thoroughly familiar with the scope and limitations of each method for which the individual is certified.

1. Set up and calibrate equipment.
2. Interpret and evaluate results with respect to applicable codes, standards and specifications.
3. Organize and report the results of nondestructive tests.
4. Exercise assigned responsibility for on the job training and guidance of Level I and trainee personnel.

Qualification for Level III

Level III personnel are qualified to perform the following tasks. A Level III is responsible for nondestructive testing operations to which assigned and for which certified. A Level III must also be generally familiar with appropriate nondestructive testing methods other than those for which specifically certified, as demonstrated by passing a Level III Basic examination.

1. Develop, qualify and approve procedures; establish and approve nondestructive testing methods and techniques to be used by Level I and Level II personnel.
2. Interpret and evaluate test results in terms of applicable codes, standards, specifications and procedures.
3. Assist in establishing acceptance criteria where none are available, based on a practical background in applicable materials, fabrication and product technology.
4. In the methods for which certified, be responsible for, and capable of, training and examination of Level I and Level II personnel for certification in those methods.

Challenges

The major challenge facing nondestructive testing personnel is to learn all that can possibly be learned during the qualification processes. Another challenge involves developing the mindset that there is something else to learn each time the nondestructive testing method is used. There is no substitute for knowledge, and nondestructive testing personnel must be demanding of themselves.

The work performed in the nondestructive testing field deserves the very best because of the direct effect of protecting life or endangering life. No field of endeavor deserves that commitment more than this one – the field of nondestructive testing.

CERTIFICATION

It is important to understand the difference between two terms that are often confused within the field of nondestructive testing: *qualification* and *certification*. Qualification is a process that should take place before a person can become certified.

According to *Recommended Practice No. SNT-TC-1A*, the qualification process for any nondestructive testing method should involve the following.

1. Training in the fundamental principles and applications of the method.
2. Experience in the application of the method under the guidance of a certified individual (on the job training).
3. Demonstrated ability to pass written and practical (hands on) tests that prove a comprehensive understanding of the method and an ability to perform actual tests using the specific nondestructive testing method.
4. The ability to pass a vision test for visual acuity and color perception or shades of gray, as needed for the method.

The actual certification of a person in nondestructive testing to a Level I, Level II or Level III is written testimony that the individual has been properly qualified. It should contain the name of the individual being certified, identification of the method and level of certification, the date and the name of the person issuing the certification. Certification is meant to document the actual qualification of the individual.

Proper qualification and certification is extremely important because the process of testing performed by certified nondestructive testing personnel can have a direct impact on the health and safety of every person who will work on, in, or in proximity to the equipment or assemblies being tested. Poor work performed by unqualified personnel can cost lives.

Modern fabrication and manufacturing projects challenge the strength and endurance of the joining techniques (such as welding) and the materials of construction. Preventive maintenance activities also present a challenge to nondestructive testing personnel.

The industries that depend on nondestructive testing cannot tolerate nondestructive testing personnel who are not adequately qualified and dedicated to good performance. Too much depends on the judgments of nondestructive testing personnel made in the work performed every day.

Employee Certification

Training
Training involves an organized program developed to provide nondestructive testing personnel with the knowledge and practicle skills necessary for qualification in a specific area. This is typically performed in a classroom setting where the principles and techniques of the particular test method are learned. Online training is also available. The length of training required is stated in the employer's written practice.

Experience
Experience includes work activities accomplished in a particular test method under the supervision of a qualified and/or certified individual in that particular method. This is to include time spent setting up tests, performing calibrations, specific techniques and other related activities. Time spent in organized training programs does not count as experience. The length of experience required is stated in the employer's written practice.

Examination
Level I and Level II personnel should be given written general and specific examinations, a practical examination and a visual examination. The general examination should cover the basic principles of the applicable method. The specific examination should cover the procedures, equipment and techniques that the employee will be required to perform in their job assignment. The practical (hands on) examination allows employees to demonstrate their ability to operate the appropriate test equipment, and to perform tests using that equipment in accordance with appropriate procedures. Level III personnel must pass written basic, method and specific examinations. Testing requirements are stated in the employer's written practice.

Certification
Certification of nondestructive testing personnel is the responsibility of the employer. Personnel may be certified when they have completed the initial training, experience and examination requirements described in the employer's written practice. The length of certification is stated in the employer's written practice. All applicants should have documentation that states their qualifications according to the requirements of the written practice before certifications are issued.

Central Certification
The American Society for Nondestructive Testing developed the ASNT Central Certification Program (ACCP) for the third party certification of nondestructive testing personnel in the U.S. This

certification program is intended to meet or exceed the requirements of *ISO-9712*.

ACCP Level II personnel must complete a written instruction examination in addition to the general, specific and practical examinations used in most other qualification examinations for certification.

ACCP Professional Level III certification requires a procedure preparation examination in addition to the basic and method examinations required for an ASNT NDT Level III certification. Additionally, a generic, hands on practical examination is a prerequisite for ACCP Professional Level III certification.

ACCP certification is transportable and valid for five years in all methods. Other certification bodies have similar central certification programs.

The employer's written practice may accept such third party certificates as proof of qualification, but the employer must still certify their nondestructive testing personnel to perform nondestructive testing.

Chapter 2

Radiographic Testing Principles

PENETRATION AND DIFFERENTIAL ABSORPTION

X-rays and gamma rays have the ability to penetrate materials, including materials that do not transmit light. In passing through material, some of these rays are absorbed. Depending on the thickness and density of the material and the size of the source being used, the amount of radiation that is transmitted through the test object will vary. The radiation transmitted through the test object produces the radiographic image.

Figure 2.1 illustrates the absorption characteristics of radiation as used in the radiographic process. The test object absorbs radiation, but less absorption occurs where the test object is thin or where a void is present. Thicker portions of the test object or dense inclusions will appear lighter because of more absorption of the radiation. Reinforcement on the outside or the inside of a weld will appear as lighter images on the radiograph because of the increased thickness and thus greater absorption. The terms *lighter* and *darker* apply to film. However, images provided by digital radiographic techniques will usually be inverted compared to film images.

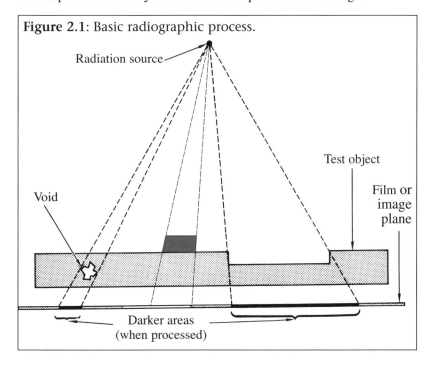

Figure 2.1: Basic radiographic process.

Radiation source

Void

Test object

Film or image plane

Darker areas (when processed)

GEOMETRIC EXPOSURE PRINCIPLES

To create a radiograph there must be a source of radiation, a test object and film or some other type of imaging detector. A radiograph is a shadow picture of a test object placed between the film/detector and the X-radiation or gamma radiation source. If the film/detector is placed too far from the test object, the discontinuity image will be enlarged. If the test object has a discontinuity and is too close to the source, the image will be greatly enlarged, resulting in the loss of dimensional acuity. Proper placement of the film/detector minimizes this enlargement and allows for more accurate representation of the size of the discontinuity.

The degree of enlargement will vary according to the relative distances of the object from the film/detector. A certain degree of enlargement will exist in every radiograph because some test objects will always be farther from the film/detector than others. The greatest enlargement is found when radiographic test objects are located at the greatest distance from the recording surface.

Enlargement cannot be eliminated entirely. Using an appropriate source-to-film/detector distance, enlargement can be minimized to a point where it will not be objectionable.

Figure 2.2 is a diagram of a radiographic exposure showing basic geometric relationships between the radiation source, the test object and the film/detector on which the image is recorded. These relationships are caused by X-rays and gamma rays obeying the laws of light. The ratio of the test object diameter D_0 to the image diameter D_f is equal to the ratio of the source-to-object distance d_0 to the source-to-film/detector distance df. For the radiographic image to be closer to the same size as the test object, the film/detector must be

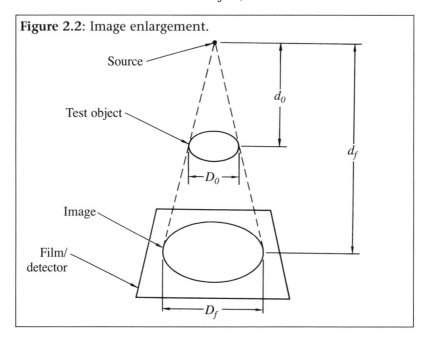

Figure 2.2: Image enlargement.

placed as close to the test object as possible, and the radiation source must be placed as far from the film or detector as is practical.

Film/Detector Image Sharpness

The sharpness of the image is determined by the size of the radiation source and the ratio of the object-to-film/detector distance and source-to-object distance. Figure 2.3a shows a small geometrical unsharpness (penumbra) when the test object is close to the film/detector. The umbra (darkest part of the shadow) is the only part that is normally seen in a radiograph. The penumbra (unsharpness) is seldom seen.

Figure 2.3b shows greater geometrical unsharpness when the source-to-film/detector distance remains unchanged but the object-to-film/detector distance is increased. Figure 2.3c shows a smaller geometrical unsharpness when the object-to-film/detector distance is the same as in Figure 2.3a, but the source-to-film/detector distance is increased.

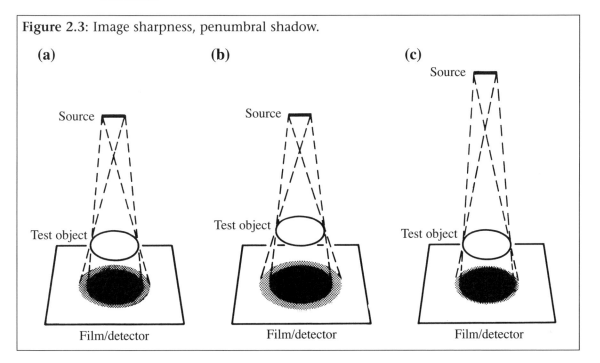

Figure 2.3: Image sharpness, penumbral shadow.

(a) (b) (c)

Most codes recommend maximum values for geometric unsharpness. To determine the geometric unsharpness, use the formula shown in Eq. 2.1.

Eq. 2.1 $U_g = \dfrac{Fd}{D}$

where U_g is geometric unsharpness, F is source size (the maximum projected dimension of the radiating source, or effective focal spot, in the plane perpendicular to the surface of the weld or object being radiographed), D is the distance from the source of radiation to weld or object being radiographed, and d is the distance from the source side of the weld or object being radiographed to the film/detector.

Optimum geometrical sharpness of the image is obtained when the radiation source is small, the distance from the source to the test object is relatively great and the distance from the test object to the film/detector plane is small. Figure 2.4 illustrates the decrease in geometrical unsharpness with a decrease in source size.

Two possible causes of film image distortion are shown in Figure 2.5. If the plane of the test object and the film/detector plane are not parallel, image distortion will result. Image distortion will also result if the radiation beam is not directed perpendicular to the film/detector plane.

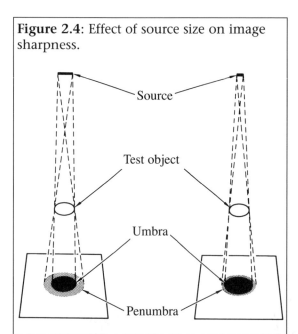

Figure 2.4: Effect of source size on image sharpness.

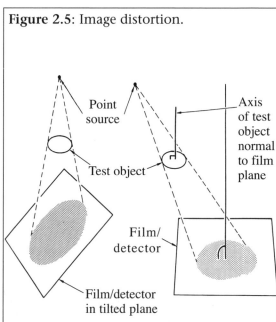

Figure 2.5: Image distortion.

X-RADIATION AND GAMMA RADIATION

As shown in Figure 2.6, X-rays and gamma rays comprise the high energy, short wavelength portion of the electromagnetic spectrum. Gamma rays and X-rays of the same wavelength have identical properties.

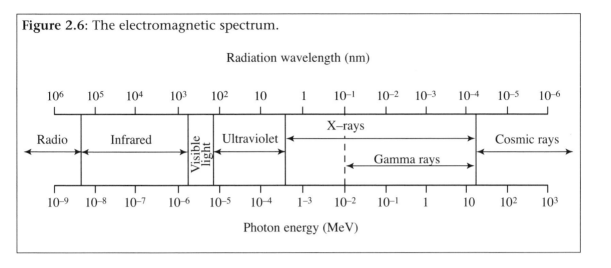

Figure 2.6: The electromagnetic spectrum.

Radiant Energy Characteristics

Radiographic testing is based on the following characteristics of X-rays and gamma rays.

1. X-rays and gamma rays have a wavelength inversely proportional to their energy.
2. X-rays and gamma rays have no electrical charge and no rest mass.
3. In free space, X-rays and gamma rays travel in straight lines at the velocity of light.
4. X-rays and gamma rays can penetrate matter, the depth of penetration being dependent on the wavelength of the radiation and the nature of the matter being penetrated.
5. X-rays and gamma rays are absorbed by matter, the percentage of absorption being a function of the matter density and thickness, and the wavelength of the radiation.
6. X-rays and gamma rays are scattered by matter, the amount of scatter being a function of the matter density and the wavelength of the radiation.
7. X-rays and gamma rays can ionize matter.
8. X-rays and gamma rays can expose film/detector by ionization.
9. X-rays and gamma rays can produce fluorescence in certain materials.
10. X-rays and gamma rays are not detectable by human senses.

X-Rays

X-rays are generated when rapidly moving (high energy) electrons interact with matter. When an electron of sufficient energy interacts with an orbital electron of an atom, a characteristic X-ray may be generated. It is called a *characteristic X-ray* because its energy is determined by the characteristic composition of the disturbed atom.

When electrons of sufficient energy interact with the nuclei of atoms, bremsstrahlung radiation (also known as *white radiation* or *braking radiation*) is generated. Bremsstrahlung radiation is also called *continuous radiation* because the energy spectrum is continuous and not entirely dependent on the disturbed atoms' characteristics.

To create the conditions required for the generation of X-rays, there must be a source of electrons, a target for the electrons to strike and a means of speeding the electrons in the desired direction. Energies of the electrons and X-rays are usually given in kilo-electron volts (keV) or million-electron volts (MeV).

The unit kilo-electron volt corresponds to the amount of kinetic energy that an electron would gain when moving between two points that differ in voltage by 1 kV. An electron would gain 1 MeV of kinetic energy when moving between two points that differ by 1 MV. The points of differing voltage are then called the *cathode* (negative) and the *anode* (positive).

Electron Source

All matter is composed, in part, of negatively charged particles called *electrons*. When a suitable material is heated, some of its electrons become agitated and escape the material as free electrons. These free electrons will surround the material as an electron cloud.

In an X-ray tube, the source of electrons is known as the *cathode*. A coil of wire (the filament) is contained in the cathode and functions as the electron emitter. When an amperage is applied across the filament, the resultant current flow heats it to electron emission temperatures.

Electron Target

X-rays are generated whenever high velocity electrons collide with any form of matter; whether solid, liquid or gas. Because the atomic number of an element indicates its density, the higher the atomic number of the chosen target material, the greater the efficiency of X-ray generation. The greater the density of the material, the greater the number of X-ray generating collisions.

In practical applications of X-ray generation, a solid material of high atomic number, usually tungsten, is used for the target. In an X-ray tube, the target is a portion of the tube anode, as shown in Figure 2.7.

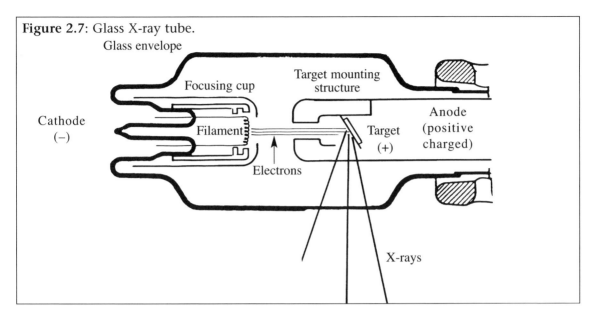

Figure 2.7: Glass X-ray tube.

Glass envelope

Focusing cup

Target mounting structure

Cathode (–)

Filament

Anode (positive charged)

Target (+)

Electrons

X-rays

Electron Acceleration

The electrons emitted at the cathode of an X-ray tube are negatively charged. Following the fundamental laws of electrical behavior, they are repelled by negatively charged objects and attracted to positively charged objects. By placing a positive charge on the anode of an X-ray tube and a negative charge on the cathode, free electrons are accelerated from the cathode to the anode. All conventional X-ray tubes use this basic principle. The X-ray tube is equipped with an internal vacuum.

X-ray tubes and associated equipment and electrical circuits are designed in many different forms determined by the need of repelling the electrons from the cathode, attracting them to the anode and accelerating them in their path.

Intensity

The number of X-rays created by electrons striking the target is one measure of the intensity of the X-ray beam. Intensity is, therefore, dependent on the amount of electrons available at the X-ray tube cathode. If all other factors remain constant, an increase in the current through the tube filament will increase the cathode temperature, cause emission of more electrons and thereby increase the intensity of the X-ray beam.

Similarly, though to a lesser degree, an increase in the positive voltage applied to the tube anode will increase the beam intensity because more of the electrons available at the cathode will be attracted to, and collide with, the target. Because the intensity of the generated beam is almost directly proportional to the flow of electrons through the tube, the output rating of an X-ray machine is often expressed in volts (kV or MeV). This same direct proportion establishes tube current as one of the exposure constants of an X-ray radiograph.

Inverse Square Law

The intensity of an X-ray beam varies inversely with the square of the distance from the radiation source. X-rays, like visible light rays, diverge on emission from their source and cover increasingly larger areas as the distance from the source increases. This relationship, illustrated in Figure 2.8, is known as the *inverse square law*. It is a major consideration in computing radiographic exposures and safety procedures.

Figure 2.8: Diagram of the inverse square law.

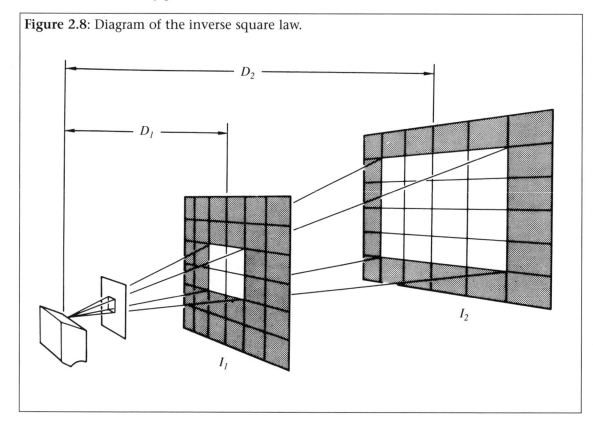

Simply put, when the distance from a known source of radiation is doubled, the intensity is one fourth as great. Conversely, if the distance to the radiation source is cut in half, the intensity is four times greater. Equation 2.1 shows one example of the inverse square law.

Eq. 2.1 $\quad \dfrac{I_1}{I_2} = \dfrac{D_2^2}{D_1^2}$

where I_1 and I_2 are the intensities at distances D_1 and D_2.

X-Ray Quality Characteristics

Radiation from an X-ray tube consists of the previously mentioned characteristic rays and continuous rays. The characteristic rays have specific wavelengths determined by the target material.

The spectrum of continuous rays covers a wide band of wavelengths and is of generally higher energy content, as shown in Figure 2.9.

Continuous rays are most commonly used in radiography. Because the wavelength of any one X-ray is partially determined by the energy (velocity) of the electron whose collision with the target caused the ray, an increase in applied voltage will produce X-rays of shorter wavelength (more energy).

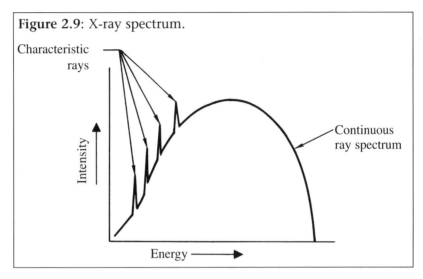

Figure 2.9: X-ray spectrum.

Figure 2.10 illustrates the effect of a change in applied voltage on the X-ray beam. An increase in applied voltage increases the intensity (quantity of X-rays) as shown, but of more importance to the radiographer is the generation of the higher energy rays with greater penetrating power. High energy (short wavelength) X-rays are known as *hard X-rays*, and low energy (longer wavelength) X-rays are known as *soft X-rays*.

Figure 2.11 illustrates the effect of a change in tube current on an X-ray beam. Variation in tube current varies the intensity of the beam, but the spectrum of wavelengths produced remains unchanged. Table 2.1 shows the intensity/hard/soft/X-ray relationships to variations in tube current and applied voltage.

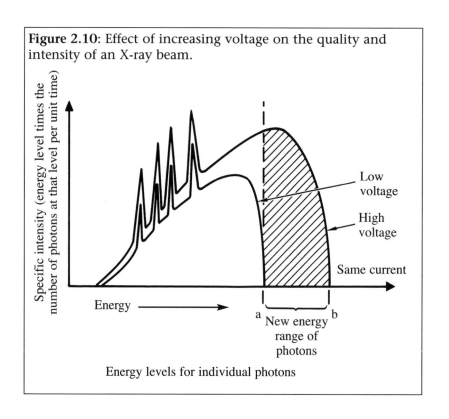

Figure 2.10: Effect of increasing voltage on the quality and intensity of an X-ray beam.

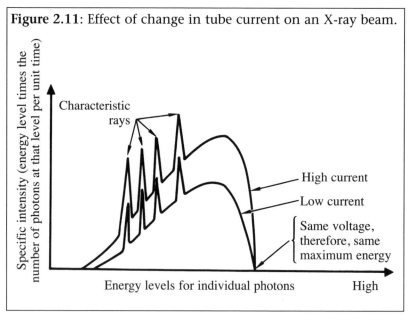

Figure 2.11: Effect of change in tube current on an X-ray beam.

Table 2.1: Effects of kilovoltage and amperage.

	Low amperage	High amperage
Low kilovoltage	Low intensity soft X-rays	High intensity soft X-rays
High kilovoltage	Low intensity hard X-rays	High intensity hard X-rays

INTERACTION WITH MATTER

To appreciate the interaction of X-rays with matter, it is necessary to consider the properties of matter that make the interaction possible. Matter is composed of numerous tiny particles called *atoms*. Atoms are made up of a dense, central, positively charged nucleus surrounded by a system of negatively charged electrons. The atom, once considered to be the smallest particle of matter, is now known to be composed of even smaller particles. The fundamental particles of interest in radiography are shown in Table 2.2.

Table 2.2: Fundamental atomic particles.

Particle	Description
Proton	A particle carrying a unit positive electrical charge. Its mass is about one atomic mass unit.
Neutron	A particle, electrically neutral, having about the same mass, but slightly heavier than, the proton.
Electron	A particle carrying a unit negative electrical charge. Its mass is 1/1840 atomic mass unit.*
Positron	A particle carrying a unit positive electrical charge and having the same mass as an electron.
* The atomic mass unit (AMU) is one twelfth the mass of the C-12 atom.	

Nuclear Atomic Concept

The nuclear atomic concept conceives the atom as consisting of a small, relatively heavy nucleus (protons and neutrons) about which electrons revolve in orbit. The volume of that portion of an atom outside the nucleus is very large compared to the volume of the nucleus itself or of the individual electrons; therefore, the greatest part of any atom is empty space. The difference in atoms of different elements is the number of protons in the nucleus. Electrically, the atom is normally in balance, the number of protons in the nucleus being equal to the number of electrons in orbit.

Ionization

Any action that disrupts the electrical balance of an atom and produces ions is called *ionization*. Atoms (minus an electron) and free (not part of any atom) subatomic particles with either a positive or negative charge are called *ions*. Free electrons are negative ions, and free particles carrying positive charges (e.g., protons) are positive ions.

X-rays passing through matter will alter the electrical balance of atoms through ionization. The energy of the ray may dislodge an electron from an atom, and temporarily free an electron. The first atom (positively charged) and the electron (negatively charged) are, respectively, positive and negative ions, also known as an *ion pair*. In this manner, X-rays cause ionization in all material in their path.

X-rays are photons (bundles of energy) traveling at the speed of light. In passing through matter, X-rays lose energy to atoms by ionization processes known as *photoelectric absorption*, *compton scattering* and *pair production*.

Photoelectric Absorption

As shown in Figures 2.12 to 2.14, when X-rays (photons) of relatively low energy pass through matter, the photon energy may be transferred to an orbital electron. This phenomenon is known as *photoelectric effect* or *absorption*. Part of the energy is expended in ejecting the electron from its orbit, and the remainder gives velocity to the electron. This energy transfer is the photoelectric effect and usually takes place with low energy photons of 0.5 MeV or less. The photoelectric process absorbs all of the energy of the photon. It is this absorption effect that makes radiography possible.

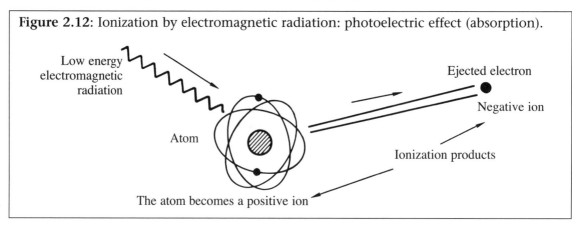

Figure 2.12: Ionization by electromagnetic radiation: photoelectric effect (absorption).

Low energy electromagnetic radiation

Ejected electron

Negative ion

Atom

Ionization products

The atom becomes a positive ion

Compton Effect

As shown in Figure 2.13, when higher energy photons (0.1 to 3 MeV) pass through matter, scattering occurs due to the compton effect. This is the term for the interaction of the photon with orbital electrons when the photon energy is not completely lost to an electron.

Part of the photon energy is expended in ejecting an orbital electron and imparting velocity to it, and the remainder, as a lower energy photon, continues onward at an angle to the original photon path. This process, progressively weakening the photon, is repeated until the photoelectric effect completely absorbs the last photon.

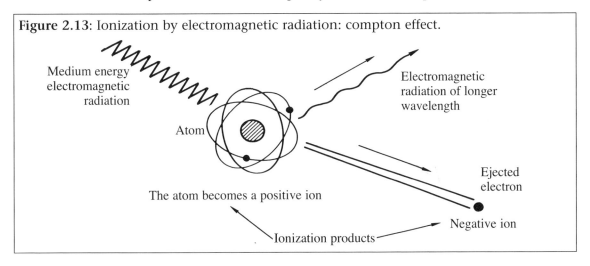

Figure 2.13: Ionization by electromagnetic radiation: compton effect.

Medium energy electromagnetic radiation

Electromagnetic radiation of longer wavelength

Atom

The atom becomes a positive ion

Ejected electron

Negative ion

Ionization products

Pair Production

Pair production occurs only with high energy photons of 1.02 MeV or more, as shown in Figure 2.14. At these energy levels, when the photon approaches the nucleus of an atom, it changes from energy to an electron-positron pair. Positrons carry a positive charge,

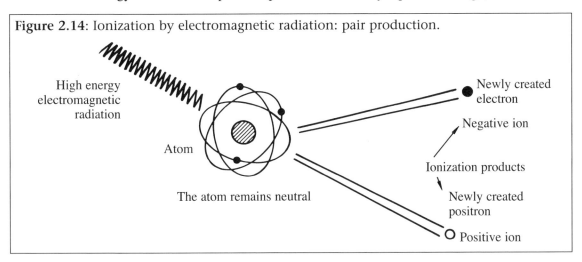

Figure 2.14: Ionization by electromagnetic radiation: pair production.

High energy electromagnetic radiation

Newly created electron

Negative ion

Atom

Ionization products

The atom remains neutral

Newly created positron

Positive ion

have the same mass as electrons and are extremely short lived. They combine at the end of their path with an electron to emit one 0.51 MeV photon subject to compton scattering and the photoelectric effect.

Scatter Radiation

The three processes (photoelectric absorption, compton scattering and pair production) all liberate electrons that move with different velocities in various directions. Because X-rays are generated whenever free electrons collide with matter, it follows that X-rays in passing through matter cause the generation of secondary X-rays. These secondary X-rays are a minor component of what is known as *scatter radiation* or *scatter*. The major component of scatter is the low energy rays represented by photons weakened in the compton scatter process. Scatter radiation is of uniformly low level energy content and of random direction.

Internal Scatter

Internal scatter is the scattering that occurs in the object being radiographed, as shown in Figure 2.15. It is reasonably uniform throughout a test object of one thickness, but affects definition by blurring the image outline.

The scatter radiation shown in Figure 2.15 obscures the edges of the test object and the hole through it. The increase in radiation passing through matter caused by scatter in the forward direction is known as *buildup*.

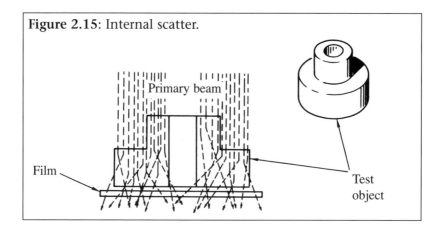

Figure 2.15: Internal scatter.

Sidescatter

Sidescatter is the scattering from walls of objects in the vicinity of the test object or from portions of the test object that cause rays to enter the sides of the test object. As shown in Figure 2.16, sidescatter obscures the image outline just as internal scatter does.

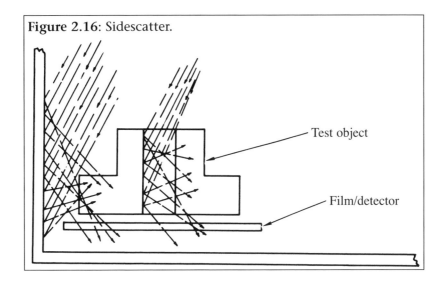

Figure 2.16: Sidescatter.

Test object

Film/detector

Backscatter

Backscatter is the scattering of rays from surfaces or objects beneath or behind the test object, as shown in Figure 2.17. Backscatter also obscures the test object image.

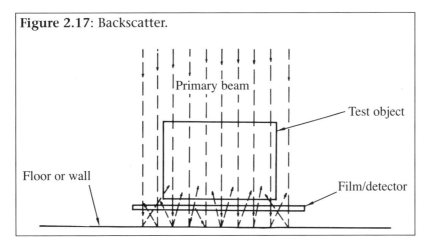

Figure 2.17: Backscatter.

Primary beam

Test object

Floor or wall

Film/detector

GAMMA RAYS

Gamma rays are produced by the nuclei of radioactive isotopes undergoing disintegration because of their basic instability. Isotopes are varieties of the same chemical element having different atomic weights. A parent element and its isotopes all have an identical number of protons in their nuclei, but a different number of neutrons. Among the known elements, there are more than 800 isotopes of which more than 500 are radioactive. The wavelength and intensity of gamma waves are determined by the source isotope characteristics and cannot be controlled or changed.

Natural Isotope Sources

Every element whose atomic number is greater than 83 has a nucleus that will probably disintegrate because of its inherent instability. Radium, the best known and most used natural radioactive source, is typical of all radioactive substances. Radium and its daughter products release energy in the following forms.

1. Gamma rays: short wavelength electromagnetic radiation of nuclear origin.
2. Alpha particles: positively charged particle having mass and charge equal in magnitude to that of a helium nuclei consisting of two protons and two neutrons.
3. Beta particles: negatively charged particles having mass and charge equal in magnitude to those of the electron.

The penetrating power of alpha and beta particles is relatively negligible; it is the gamma rays that are of use to the radiographer.

Artificial Sources

There are two sources of manufactured radioactive isotopes (radioisotopes). Atomic reactor operation involving the fission of U-235 results in the production of many different isotopes usable as radiation sources.

One of the radioisotopes used in radiography is cesium-137 (Cs-137). It is obtained as a by product of nuclear fission. The second, and most common, means of creating radioisotopes is by bombarding certain elements with neutrons. The nuclei of the bombarded element are changed, usually by the capture of neutrons; and, thereby, may become unstable or radioactive.

Commonly used radioisotopes obtained by neutron bombardment are cobalt-60 (Co-60), thulium-170 (Tm-170), selenium-75 (Se-75) and iridium-192 (Ir-192). The numerical designator of each of these radioisotopes denotes its mass number and distinguishes it from the parent isotope and other isotopes of the same element. Artificially produced isotopes emit gamma rays, alpha particles and beta particles in exactly the same manner that natural isotopes do.

Gamma Ray Intensity

Gamma ray intensity is often measured in roentgens per hour or sieverts per hour at one foot; a measure of radiation emission over a given period of time at a fixed distance. The activity (amount of radioactive material) of a gamma ray source determines the intensity of its radiation. The activity of artificial radioisotope sources is determined by the effectiveness of the neutron bombardment that created the isotopes. The measure of activity is the curie (3.7×10^{10} disintegrations per second or 37 gigabecquerel).

Specific Activity

Specific activity is defined as the degree of concentration of radioactive material within a gamma ray source. It is usually expressed in terms of curies per gram or curies per cubic centimeter. Two isotope sources of the same material with the same activity (curies) having different specific activities will have different dimensions. The source with the greater specific activity will be the smaller of the two. For radiographic purposes, specific activity is an important measure of radioisotopes because the smaller the radioactive source, the greater the sharpness of the resultant image, as shown in Figure 2.4.

Half Life

The length of time required for the activity of a radioisotope to decay (disintegrate) to one half of its initial strength is called its *half life*. The half life of a radioisotope is a basic characteristic, and is dependent on the particular isotope of a given element. In radiography, the half life of a gamma ray source is used as a measure of activity in relation to time. Dated decay curves similar to that shown in Figure 2.18 are supplied when the radioisotopes are obtained.

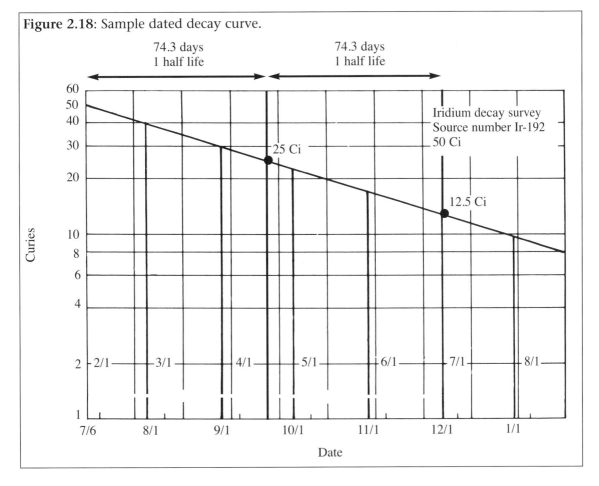

Figure 2.18: Sample dated decay curve.

Inverse Square Law

Gamma rays and X-rays have identical propagation characteristics because they both conform to the laws of light. Just as it does with X-rays, the intensity of gamma ray emission varies inversely with the square of the distance from the source, as was shown in Figure 2.8. Remember that when the distance from a known intensity is doubled, the exposure will be one fourth as great.

Gamma Ray Quality Characteristics

Radiation from a gamma ray source consists of rays whose wavelengths (energy) are determined by the nature of the source. Each of the commonly used radioisotopes has specific uses because of its fixed gamma energy characteristics. Table 2.3 lists the most used radioisotopes and the energy of their gamma ray emissions. Note that several radioisotopes have multiple wavelengths.

Table 2.3: Equivalent gamma ray energy.	
Radioisotope	**Equivalent gamma ray energy (MeV)**
Cobalt-60	1.33 1.17
Iridium-192	0.31 0.47 0.6
Thulium-170	0.084 0.052
Cesium-137	0.66
Selenium-75	0.09 to 0.4

Gamma Ray Interaction with Matter

The ionization, absorption, scattering and pair production caused by gamma ray interaction with matter are identical to those of X-rays.

Chapter 3

Equipment

Introduction

Radiographic equipment as discussed in this chapter is limited to radiation source equipment that generates either X-radiation or gamma radiation. Additional equipment required to produce a radiograph or other visual representation of a test object is discussed in later chapters.

X-Ray Equipment

The three basic requirements for the generation of X-rays are a source of free electrons, a means of moving the electrons rapidly in the desired direction and a suitable material for the electrons to strike. The design of modern X-ray equipment is a result of refinements in the methods of satisfying these requirements, and focus the X-rays in a useful manner. X-ray equipment is usually regulated by state agencies.

Portable X-Ray Units

Portable X-ray units are very important in field radiography. They are used on pipelines, bridges, vessels, etc. The X-ray units incorporate special components, such as integrated tube heads and tube inserts. Portable X-ray units are light weight and compact, making them ideal for field inspection. Many of the portable units available today are air-cooled, thus eliminating the need for water or oil accessories.

X-Ray Tube

The productive portion of X-ray equipment is the tube. The remaining components of an X-ray machine are designed to support the function of the tube or to meet safety requirements. The tube consists of two electrodes, the cathode and the anode, enclosed in a high vacuum envelope of heat resistant glass or ceramic. The filament portion of the cathode functions as a source of free electrons, and the anode houses the target which the electrons strike.

Associated with the tube is equipment that heats the filament, speeds and controls the resultant free electrons in a path to the anode, removes the heat generated by the X-ray generation process and shields the equipment and surrounding area from unwanted radiation. As shown in Figure 3.1, there are many variations in the size and shape of X-ray tubes.

Figure 3.1: X-ray tubes.

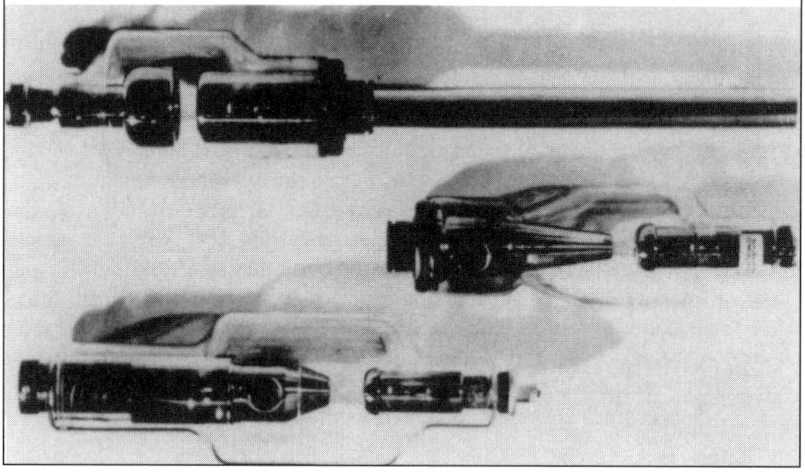

Tube Envelope

The tube envelope is constructed of glass or ceramic that has a high melting point because of the extreme heat generated at the anode. Structurally, the envelope has sufficient strength to resist the implosive force of the high vacuum interior.

The shape of the envelope is determined by the electrical circuitry used with the tube and the desired tube use. Electrical connections through the envelope to the tube's electrodes are made in either of two ways. They are made through insulation material able to withstand the temperature, pressure and electrical forces of the X-ray generating process, or by connection to the envelope itself. Electrical connections to the envelope are accomplished with metal alloys that have a coefficient of thermal expansion similar to that of the glass or ceramic. The alloy is fused with the glass or ceramic and becomes part of the envelope. A high vacuum environment for the tube elements is necessary for the following reasons.

1. Prevents oxidation of the electrode materials.
2. Permits ready passage of the electron beam without ionization of gas within the tube.
3. Provides electrical insulation between the electrodes.

Cathode

The cathode of the X-ray tube incorporates a focusing cup and the filament. Usually constructed of very pure iron and nickel, the focusing cup functions as an electrostatic lens whose purpose is to direct the electrons in a beam toward the anode.

The electron emitting portion of the cathode is the filament, which is brought to the required high temperature by the flow of electrical current through it. The filament is usually a coil of tungsten wire, because tungsten has the desired electrical and thermal characteristics. The placement of the filament within the focusing cup and the shape of the cup determine the dimensions of the electron beam and the resultant area of X-ray emission at the target.

Filament Heating

Because of the electrical characteristics of tungsten, a small flow of current through the filament suffices to heat it to temperatures that cause electron emission. Any change in the amperage applied to the filament varies the filament current and the number of electrons emitted. A change in the number of emitted electrons varies with the electron flow (current) through the tube. On most X-ray machines, tube current control is obtained by regulating amperage applied across the filament. As the tube current, measured in milliamperes (mA) increases, the intensity of the X-rays increases but the wavelength remains the same.

Anode

The anode of the X-ray tube is a metallic electrode of high electrical and thermal conductivity. Usually it is made of copper with that portion directly facing the cathode being tungsten, gold or platinum. It is these latter materials that function as the target.

Copper and tungsten are the most common anode materials because copper has the necessary electrical and thermal characteristics, and tungsten is an economically feasible, dense material with a high melting point. A dense target material is required to ensure a maximum number of collisions when the electron beam strikes the target. Material with a high melting point is necessary to withstand the heat of X-ray generation.

Focal Spot

The sharpness of a radiographic image is partly determined by the size of the radiation source (focal spot). The electron beam in most X-ray tubes is focused so that a rectangular area of the target is bombarded by the beam, called a *target*. Usually the anode target is set at an angle, as shown in Figure 3.2, and the projected size of the bombarded area, as viewed from the test object, appears smaller than the actual focal spot. This projected area of the electron beam is the effective focal spot.

In theory, the optimum tube would contain a pinpoint focal spot. In practice, the size to which the focal spot can be reduced is limited by the heat generated in target bombardment. If the focal spot is reduced beyond certain limits, the heat at the point of impact destroys the target.

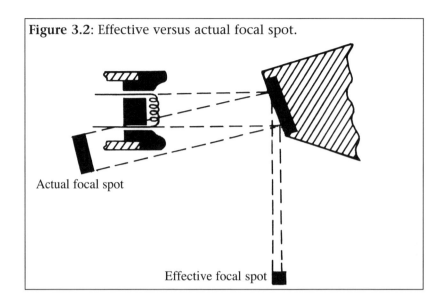

Figure 3.2: Effective versus actual focal spot.

Actual focal spot

Effective focal spot

Linear Accelerators

Linear accelerators are the simplest type of accelerators. Linear accelerators consist of a long line of coils, which charged particles are accelerated through.

There are two types of linear accelerators. One type is the standing wave linear accelerator, in which particles travel along a cylindrical vacuum tank through a series of drift tubes separated by gaps. As the particles cross the gaps, electromagnetic waves, called *standing waves*, accelerate them. If the current was kept, it would pull the particle back toward the tube when it leaves. The waves provide an electric field that speeds up the particles by acting on their electric charges. This type of accelerator can only manage to accelerate particles to 200 MeV (million electron volts).

The other type of linear accelerator is the traveling wave linear accelerator. This accelerator speeds particles through a single long pipe by an electromagnetic wave that travels with the particle. This high frequency wave is called a *travelling wave*. As long as the wave speed matches the particles' speed, the particles will continue to gain energy. This type of accelerator can accelerate particle to 30 GeV (giga-electron volts, or billion electron volts). The general arrangement of a linear accelerator is shown in Figure 3.3.

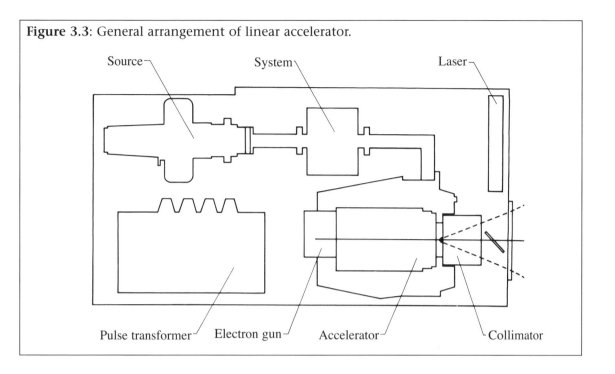

Figure 3.3: General arrangement of linear accelerator.

Source

System

Laser

Pulse transformer

Electron gun

Accelerator

Collimator

X-Ray Beam Configuration

X-rays are radiated in all directions from the tube target. Once created, they cannot be focused or otherwise directed. The direction of useful X-radiation is determined by the target positioning at the tube anode and the placement of lead shielding about the tube. With selected positioning of the target and variations in shielding placement, almost any beam configuration can be obtained, as shown in Figure 3.4.

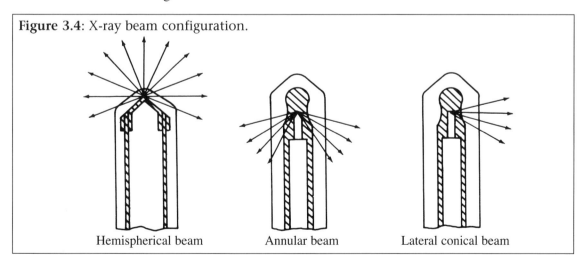

Figure 3.4: X-ray beam configuration.

Hemispherical beam

Annular beam

Lateral conical beam

Accelerating Potential

The operating voltage (difference in electrical potential between the cathode and anode) applied to an X-ray tube determines the penetrating effect of X-radiation, as was shown in Figure 2.10. The higher the voltage, the greater the electron velocity and the shorter the wavelengths of the generated X-rays. In other words, as kilovoltage increases, wavelength shortens and penetration increases. The high voltage necessary to generate short waves of great penetrating power is obtained from transformers, electrostatic generators or accelerators.

Iron Core Transformers

The majority of X-ray equipment used in industrial radiography uses iron core transformers to produce required high voltages. A typical self rectifying, high voltage circuit for X-ray equipment is shown in Figure 3.5. The basic limiting factors to iron core transformer use are their size and weight. Iron core transformers are used to produce voltages up to 400 kVp, usually in self rectified circuits; however, they are often used with half wave and full wave rectifiers, voltage doublers and constant potential circuits. Iron core

Figure 3.5: Standard high voltage circuit designs for portable tank units: (a) cathode grounded; (b) center grounded; (c) anode grounded.

(a)

(b)

(c)

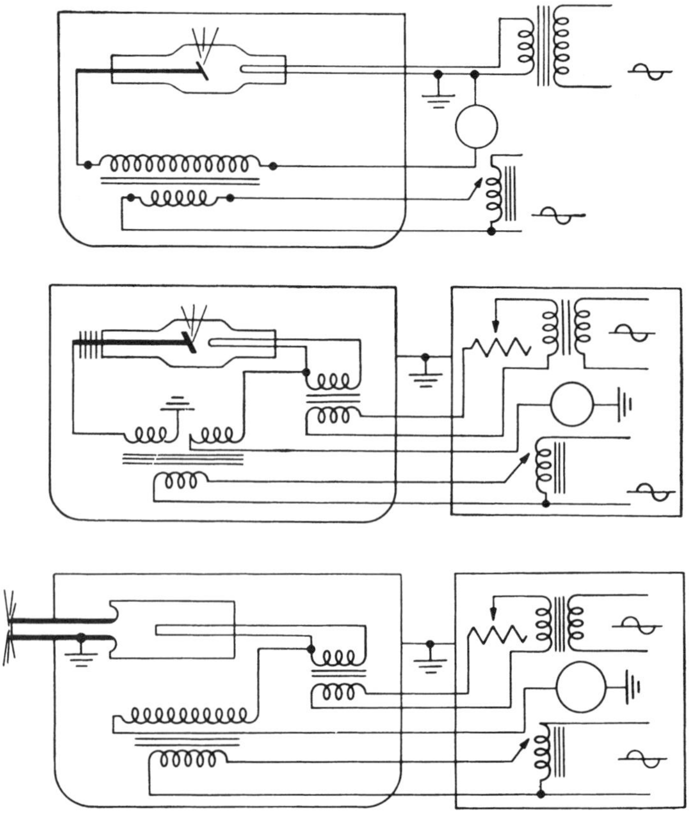

transformers in modern X-ray equipment are either mounted in tubehead tank units with the tube, or are separately housed.

Heat Dissipation

The process of X-ray generation is very inefficient, and most of the energy of the electron beam in the tube is expended in the production of heat. To avoid destruction of the tube anode, this heat must be dissipated. Heat dissipation in medium and low power equipment is usually accomplished through an external finned radiator that is in good thermal connection with the anode and is cooled by a flow of oil or gas about its surfaces.

Higher power equipment makes use of injection cooling. The coolant, oil or water is circulated through the hollow anode of the tube. Because the duty cycle (percentage of exposure time versus total time that the equipment is running) of X-ray equipment is determined by the rate of anode cooling, the efficiency of equipment cannot exceed the efficiency of its cooling system.

Equipment Shielding

X-radiation can be controlled by shielding. X-ray tubes, or the tubeheads in which they are contained, are shielded by lead plates or sleeves to prevent unwanted radiation spread. The design of this shielding varies with equipment, but in all cases it serves to absorb that portion of the radiation that is not traveling in the desired direction. In any X-ray equipment, the angle of coverage of the X-ray beam is a function of the target angle, the geometry of the focal spot position and the X-ray port size as determined by shielding placement.

Tubeheads

Tubeheads, shown in Figure 3.6, used with portable X-ray equipment consist of an outer metallic shell with an X-ray port and usually contain the X-ray tube, high voltage and filament transformers, insulating oil or gas and lead shielding. Tubeheads used with permanently installed X-ray equipment contain all of the foregoing items except the transformers, which are housed in a separate unit.

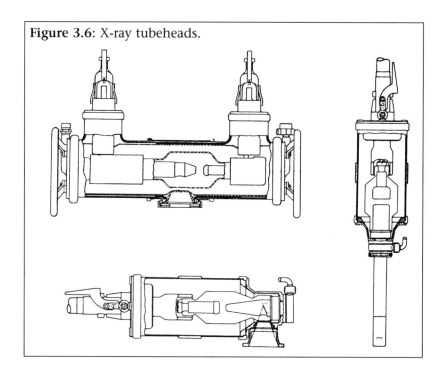

Figure 3.6: X-ray tubeheads.

CONTROL PANEL

The control panel of an X-ray machine is designed to permit the technician to control the generation of X-rays so that exposures can be made simply and rapidly. The panel also provides protective electrical circuits that prevent damage to the equipment. Dependent on the complexity of the equipment and the electrical circuitry design, the control panel will be comprised of some, or all, of the following controls and indicators.

1. **Line voltage selector switch**: Permits equipment operation with various line voltages such as 110 V alternating current, 220 V alternating current, etc.
2. **Line voltage control**: Permits adjustment of line voltage to exact values.
3. **Line voltage meter**: A voltmeter indicating the line voltage used in conjunction with the line voltage control.
4. **High voltage control**: Permits adjustment of the voltage applied across the tube.
5. **High voltage meter**: A voltmeter, usually calibrated in kilovolts (kV), used in conjunction with the high voltage control.
6. **Tube current control**: Permits adjustment of tube current to exact values.

7. **Tube current meter**: An ammeter, usually calibrated in milliamperes (mA), used in conjunction with tube current control.
8. **Exposure time**: A synchronous timing device used to time exposures.
9. **Power on/off switch**: Controls the application of power to the equipment; usually applies power to the tube filament only.
10. **Power indicator lamp**: Visual indication that the equipment is energized.
11. **High voltage on/off switch**: Controls the application of power to the tube anode.
12. **High voltage indicator lamp**: Visual indication that the equipment is completely energized and X-rays are being generated.
13. **Cooling indicator lamp**: Visual indication that the cooling system is functioning.
14. **Focal spot selector control**: Used with tubes having two focal spots; permits selection of desired size focal spot.

Internally, the control panel contains protective electrical circuits.

1. **Overload circuit breaker**: Provides protection for the equipment by removing power when the equipment becomes overheated as a result of component failure.
2. **Overvoltage protection circuit**: Bleeds off excess voltage caused by surges in the line voltage supply.
3. **Overcurrent relay**: Prevents excess current flow through the tube by controlling the filament voltage.

Equipment Protective Devices

The electrical protective devices of the control panel serve to protect the equipment against electrical malfunctions. Additional protection is provided against excess heat or insulation failure.

1. **Over temperature thermostat**: Installed in the tubehead, functions to remove power from the equipment when excess heat is present.
2. **Pressure stats**: Installed in the tubehead of equipment using gas for insulation to remove power from the equipment when the gas pressure is below safe values.
3. **Flow switches**: Installed in the oil and water circulators of equipment cooled by these means to remove power from the unit when the cooling system fails.

GAMMA RAY EQUIPMENT

Gamma radiation from radioactive material cannot be shut off. Gamma ray equipment is designed to provide radiation safe storage and remote handling of a radioisotope source. The United States Nuclear Regulatory Commission (NRC) and various state agencies prescribe safety standards for the storage and handling of radioisotopes under their control.

Safety procedures are required for the storage and handling of all radioisotopes. Safety procedures are also required for storage. Information on radiation safety can be found in Chapter 5.

Gamma Ray Sources

The effective focal spot in X-radiography is the X-ray generating portion of the target as viewed from the test object. In gamma radiography, because all of the radioactive material is producing gamma rays, the focal spot is the surface area of the material as viewed from the test object. For this reason, it is desirable that the dimensions of a gamma ray source be as small as possible.

Most isotopes used in radiography are round wafers encapsulated in a stainless steel cylinder.

Radium

Radium is a natural radioactive substance having a half life of about 1600 years.

Pure radium is not used in radiography, and most sources consist of radium sulfate packaged in either spherical or cylindrical capsules. Because of its low specific activity, radium is rarely used in industrial radiography.

Cobalt-60

The artificial isotope cobalt-60 (Co-60) is created by neutron bombardment of cobalt-59 and has a half life of 5.3 years. Table 3.1 shows its decay rate of 6 month intervals during one half life cycle.

The primary gamma ray emission of Co-60 consists of 1.33 and 1.17 MeV rays similar in energy equivalency to the output of a 2 MeV X-ray machine. The radioisotope is supplied in the form of a capsuled pellet, and may be obtained in different sizes. It is used for radiography of steel, copper, brass and other medium weight metals of thicknesses ranging from 1 to 9 in. (2.5 to 23 cm). Because of its penetrating radiation, its use requires thick shielding with associated weight and handling difficulty.

Table 3.1: Decay rate for Co-60.

Time (years)	0	0.5	1	1.5	2	2.5	3	3.5	4	4.5	5	5.3
Percent Activity	100	93.6	87.7	82.2	77	72	67.5	63.3	59.2	55.5	52	50

Iridium-192

Another artificial isotope produced by neutron bombardment is iridium-192 (Ir-192). It has a half life of 74.3 days. It has high specific activity and emits gamma rays ranging from 0.31 to 0.6 MeV, comparable in penetrating power to those of a 600 kVp X-ray machine. Industrially, it is used for radiography of steel and similar metals of thicknesses between 0.25 and 3 in. (0.6 and 7.6 cm). Its relatively low energy radiation and its high specific activity combine to make it an easily shielded, strong radiation source of small physical size (focal spot). The radioisotope is obtainable in the form of encapsulated wafers.

Selenium-75

The radioisotope selenium-75 (Se-75) has a softer gamma ray spectrum than Ir-192, as well as a longer half life of about 120 days. Since Se-75 has considerably lower radiation energies equivalent to 200 kVp, it results in improved quality of weld radiographs.

Thulium-170

The radioisotope thulium-170 (Tm-170), obtained by neutron bombardment of thulium-169, has a half life of about 130 days. The disintegration of the isotope produces 84 kVp and 52 kVp gamma rays, soft rays similar to the radiation of X-ray equipment operating in the 50 to 100 kVp range.

The radioisotope Tm-170 is the best for radiography of thin metals because it is capable of producing good radiographs of steel test objects less than 0.5 in. (1.3 cm) thick. One of the major advantages of Tm-170 is its soft wave radiation, which permits its containment in small portable units, because only a small amount of shielding is required. Because the pure metal is difficult to obtain, the radioisotope is usually supplied in capsules containing the oxide Tm-203 in powder form.

Cesium-137

The radioisotope cesium-137 (Cs-137) is a powder that is considered an unstable (radioactive) isotope. It is useful in radiography because of its long half life of 30 years. Cesium-137 undergoes radioactive decay with the emission of beta particles and relatively strong gamma radiation. Because of the chemical nature of cesium, it moves easily through the environment. This makes cleanup difficult.

Cesium-137 is widely used in the construction industry for moisture-density gages, and as a calibration source for survey meters. It is also used in plants and refineries in leveling gages to detect the liquid flow in pipes and tanks. It can also be used in thickness gages for measuring the thickness of sheet metal, paper and film.

Other Radioisotopes

Many other radioisotopes that are radiographically useful are not considered here because in practical applications, one or another of the five discussed is superior. Table 3.2 is a summary of the characteristics of the most used isotopes.

Table 3.2: Radioisotope characteristics.

Radioisotope	Co-60	Ir-192	Tm-170	Cs-137	Se-75
Half life	5.3 years	74.3 days	130 days	30 years	120 days
Chemical form	Co	Ir	Tm_2O_3	CsCl	Se
Gamma rays (MeV)	1.17 1.33	0.31 0.47 0.6	0.084 0.052 0.032*	0.66 4.2	97 to 401 keV
Practical sources (curies)	100 1.3	200 0.48	50	75 0.32	75 30
Approximate maximum focal spot size	0.5 in. (1.2 cm)	0.3 in. (0.7 cm)	0.1 in. (0.3 cm)	0.4 in. (1 cm)	0.08 to 0.1 in. (0.2 to 0.3 cm)

*Varies widely because of high self absorption.

Isotope Cameras

Because of the ever present radiation hazard, isotope sources must be handled with extreme care and stored and locked in adequately shielded containers when not in use. Equipment to accomplish safe handling and storage of radioisotope sources, together with a source, is often called a *camera* or *exposure device*.

Portable industrial radiographic exposure devices are designed for field or laboratory use. They are self contained units, meaning no external power supply is required. The exposure devices contain positive, self locking mechanisms ensuring safety in accordance with ANSI and ISO requirements, in addition to NRC and IAEA requirements. All exposure devices must withstand rigorous testing before being sold in the United States.

Chapter 4

Radiographic Film

INTRODUCTION

Radiographic film consists of a thin, transparent plastic sheet, or base, coated on one or both sides with an emulsion of gelatin, about 0.001 in. (0.003 cm) thick, containing very fine grains of silver bromide (AgBr). The lightly colored gelatin is almost transparent to X-rays, and it is mainly the silver bromide that will absorb X-rays or gamma rays.

When exposed to X-rays, gamma rays or visible light rays, silver bromide crystals undergo a reaction that makes them more open to the chemical process (developing) that converts them to black metallic silver. Putting emulsion on both sides of the base doubles the amount of radiation sensitive silver compound, and therefore increases the speed. The emulsion layers are thin enough so that developing, fixing and drying can be accomplished in a reasonable time.

Exposure to radiation creates a latent (hidden) image on the film, and chemical processing makes the image visible. Because the radiation source, the test object and the conditions of exposure determine the amount of radiation reaching the film at any given point, the radiographer is primarily concerned with those film characteristics that fix the density and sharpness of the processed film image in the finished radiograph.

When X-rays or gamma rays hit the grains of the silver bromide, a change takes place in the physical structure of the grains. This change is of such a nature that it cannot be detected by ordinary physical methods.

When the exposed film is treated with a chemical solution (the developer), a reaction takes place causing the formation of black metallic silver. It is this silver, suspended in the gelatin on both sides of the base, that constitutes the image. Because the properties of X-rays and gamma rays are different from light or other forms of radiation, the emulsion must be different from those used in other types of photography.

Usefulness of Radiographs

The usefulness of any radiograph is measured by its impact on the human eye. When the radiographer interprets a radiograph, the details of the test object image are seen in terms of the amount of light passing through the processed film, as shown in Figure 4.1. Areas of high density (areas exposed to relatively large amounts of radiation) will appear dark gray; areas of light density (areas exposed to less radiation) will appear light gray. The density (darkness) difference between any two film areas is known as *radiographic contrast* (difference). The sharpness of any change in density is called *definition*. Successful interpretation of any radiograph relies on the ability of the radiographer to see the contrast and definition in the radiograph.

Figure 4.1: Typical radiographs of discontinuities.

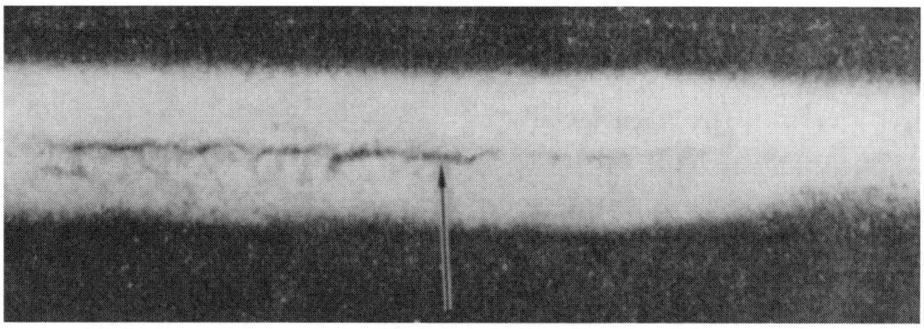

Longitudinal cracks

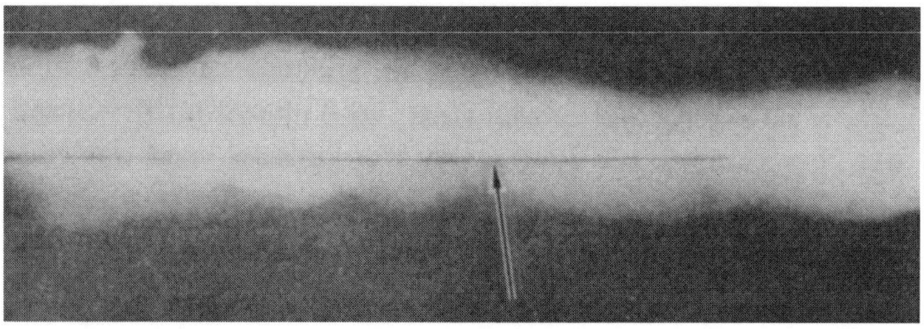

Lack of fusion

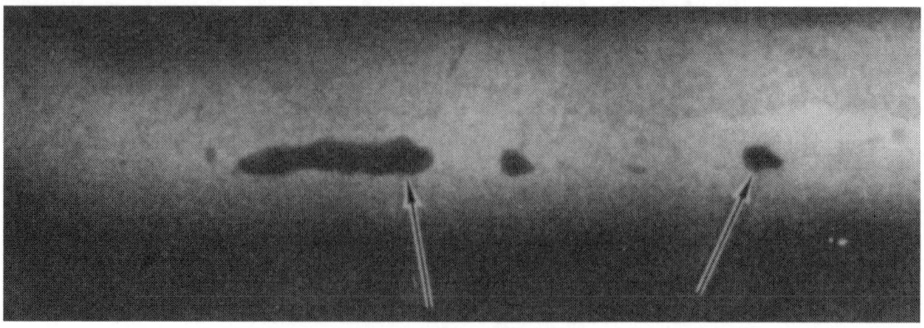

Elongated voids

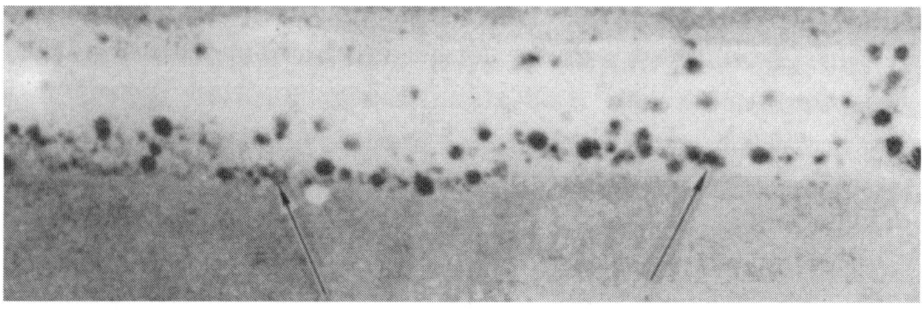

Porosity

RADIOGRAPHIC CONTRAST

Radiographic contrast, as shown in Figure 4.2, is defined as the difference in density between any two selected portions. It is a combination of subject contrast and film contrast and, for any particular test object, depends on radiation energy applied (penetrating quality), contrast characteristics of the film, exposure (the product of radiation intensity and time), screens, film processing and scattered radiation.

Figure 4.2: Radiographic contrast: (a) exposure setup; and (b) viewing setup.

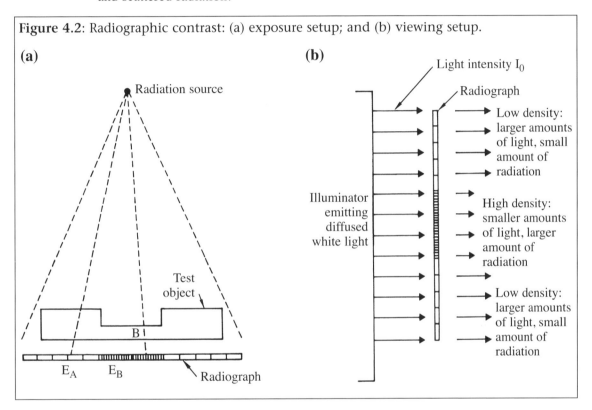

Subject Contrast

Subject contrast is the relative radiation intensities passing through any two selected portions of a material. Uniform test objects of little thickness variation have low subject contrast. Subject contrast depends on the following factors reaching the film.

1. Type or configuration of the test object.
2. Energy of the radiation used, wavelength, type of isotopic source.
3. Intensity and amount of the scattered radiation.

Subject contrast decreases as the wavelength of the incident radiation decreases, i.e. as the energy of the incident radiation increases. Large thickness variations produce high subject contrast. Subject contrast may be modified by use of different X-ray or gamma ray energies, masks, diaphragms, filters or screens.

Film Contrast

The ability of film to detect and record different radiation exposures as differences in density is called *film contrast*. Radiographic film is manufactured with a variety of emulsions that give different film contrasts and other properties, such as speed and graininess. The contrast values of any particular film are usually expressed as a relationship between film exposure and the resulting film density. The relationship is expressed in the form of film characteristic curves.

Film contrast is determined by the following factors.

1. Grain size or type of film.
2. Chemistry of the film processing chemicals.
3. Concentrations of the film processing chemicals.
4. Development time.
5. Development temperature.
6. Type of agitation.

Characteristic Curves

Exposure is defined as the product of the intensity of the radiation reaching the object, and the time the object is exposed to that radiation. Characteristic curves are graphs showing how the amount of radiation exposure on a photographic material relates to the optical density of the image. A different curve is obtained for the same film using different developers or development times. Unexposed but developed film has a low density, known as *base fog* (the fog will be higher if the film is developed for a longer time).

The output levels of X-ray equipment is directly proportional to the tube current and time; therefore, it is also directly proportional to their product. Mathematically, $E = Mt$, where E is the exposure, M is the tube current in milliamperes (mA) and t is the exposure time. It is this relationship that permits X-ray exposure, at a given kilovoltage, to be specified in terms of milliampere minutes or milliampere seconds without stating specific values of tube current or time.

Similarly, gamma ray exposure is stated as $E = Mt$, where E is the exposure, M is the source activity and t is the exposure time. Thus, gamma ray exposures may be expressed in curie-minutes, millicurie minutes or millicurie seconds without stating specific values of source activity or time.

There are no convenient units suitable to all X-radiation and gamma radiation conditions in which to express radiographic exposures. For this reason, relative exposure is used as one axis in plotting the characteristic curve for a given source.

By this means, any exposure given a film is expressed in terms of any other exposure, which produces a relative scale. For convenience, the logarithm (exponent indicating the power of the base number) of the relative exposure itself is used, because the logarithm compresses what would otherwise be a long scale.

Similar results are obtained if semilog paper is used and actual exposure is laid out on the logarithmic scale. Film density, the second axis used in plotting film characteristic curves, is laid out on a linear scale. It is the common logarithm of the ratio of light incident on one side of a radiograph to the light transmitted through the radiograph, as shown in Eq. 4.1.

Eq. 4.1 $\quad D = \log_{10} \dfrac{I_o}{I}$

where D is the film density, I_o is the intensity of the incident light and I is the intensity of the transmitted light.

It is difficult for the human eye to distinguish between small density differences. There is a lower limit of contrast that the eye cannot detect. The characteristic curves for most films make it readily apparent that as exposure increases, overall film density increases and, more importantly, film contrast increases.

In Figure 4.3, film exposure E_A is less than E_B, and it is the difference between the two that the radiograph must make clear in terms of film density. For a low exposure E_1, the difference in density between E_A and E_B is relatively small and will probably not be discernible by the eye.

By increasing the exposure to the value represented by E_2, not only is the overall density of the radiograph increased, the density difference (radiographic contrast) between E_A and E_B is greatly increased. The resulting contrast is easily detectable by the eye.

Selection of a correct exposure has used the film's contrast characteristics to amplify the subject contrast resulting in a useful radiograph. In industrial radiography, films should always be exposed for a density of at least 1.5. The highest desirable density is limited by the light intensity available for reading the radiograph. In all cases, the film densities must meet the applicable standard or code requirements. Most codes and specifications give upper and lower density limits, usually 1.8 to 4.0.

Film speed is measured by the exposure required to obtain a desired film density. High speed film needs only low exposure, whereas slow speed film requires more exposure to attain the same film density. Figure 4.4 illustrates characteristic curves for three different speed films. The shape of each curve and its position on the log relative exposure axis is determined by the design of the film. Film speed is a consideration of importance because time is a cost factor in any industrial operation. Whenever other considerations, such as acceptable graininess permit, fast film may be used.

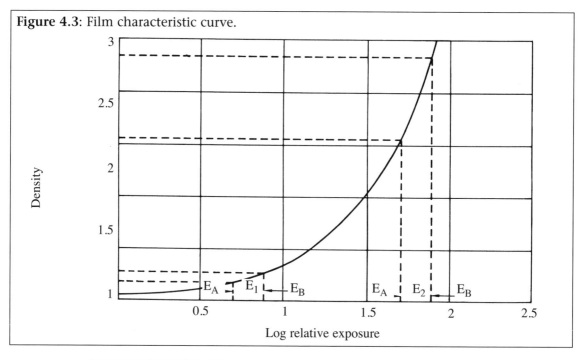

Figure 4.3: Film characteristic curve.

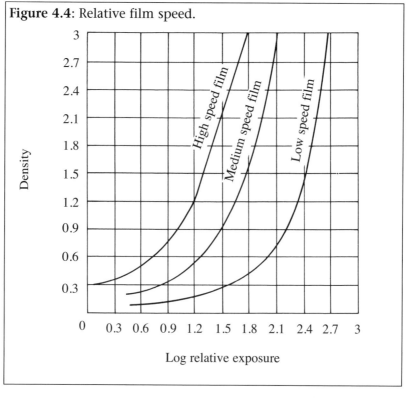

Figure 4.4: Relative film speed.

Graininess is the visible evidence of the grouping into clumps of the minute silver particles (grains) that form the image on radiographic film. It affects film contrast and image definition, and all film is subject to it, as shown in Figure 4.5. Small grains provide better definition because they can better outline small areas of film. Larger grains (fast film) may cause blurring of the outline of discontinuities. The degree of graininess of an exposed film is dependent on the following factors.

1. The fine or coarse grain structure of the film emulsion.
2. The quality of the radiation to which the film is exposed, because an increase in the penetrating quality of the radiation will cause an increase in graininess.
3. Film processing, because graininess is directly related to the development process. Under normal conditions of development, any increase in development time is accompanied by an increase in film graininess.
4. Fluorescent screens that cause increased graininess with increase in radiation energy.

Figure 4.5: Film grain variations: (a) fast film (large grains); and (b) slow film (small grains).

(a)   **(b)**

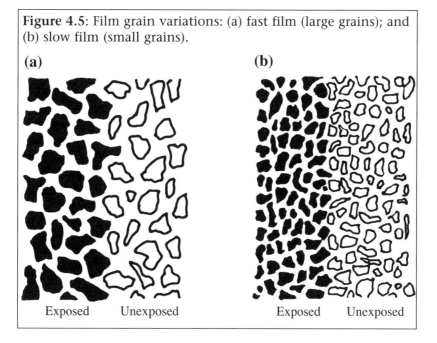

Exposed   Unexposed   Exposed   Unexposed

FILM SELECTION FACTORS

Most of the time, the selection of film is made by the radiographer. Sometimes the customer will require a particular brand and/or type. The selection of film by the radiographer is based on the need for radiographs of a certain contrast and definition quality. The radiographer must be familiar with the following factors.

1. Thickness and density of the test object.
2. The type of indication or discontinuity normally associated with the test object.
3. Size of an acceptable indication.
4. Accessibility and location and configuration of the test object.
5. Customer requirements.

Film contrast, speed and graininess are interrelated. Fast films usually have large grains and poor resolution/sensitivity, although the exposure time is shorter. Slow films have fine grain and good resolution/sensitivity. The radiographer and the customer must realize that the finer the grain of the film, the longer the exposure will be, however, the sensitivity will be enhanced. Therefore, though it is economically advantageous to make exposures as short as possible, fast film is limited by the graininess that can be tolerated in the radiograph.

Manufacturers have created films of various characteristics, each designed for a specific purpose. Their recommendations for film usage are reliable.

Film comes in lightproof containers, and all film (prior to being developed) must be protected from light. Even momentary exposure to light, other than a safelight) will ruin film.

FILM PROCESSING

Once a radiographic exposure has been made, the film must be processed so that the latent image produced by the radiation becomes visible. All of the procedures involved in making a radiograph are important, because processing errors can make an otherwise worthwhile radiograph useless. Each step in film processing is dependent on the step preceding it and, in turn, affects those following.

Processing Precautions

To obtain consistently good results, the following general precautions must be observed in processing radiographic film.

1. Maintain chemical concentrations, solution temperatures and processing times within prescribed limits.
2. Use equipment, tanks, trays and holders that can withstand the chemical action of the processing solutions without contaminating the solutions.
3. Ensure tanks are cleaned and rinsed thoroughly before filling with chemicals. The stick used to stir chemicals should be made of stainless steel to avoid contamination.
4. Equip the darkroom with suitable safeguards and lighting controls to avoid fogging film. Periodically, safelight filters should be checked for light leaks.

5. Maintain cleanliness, especially in the darkroom. Lint and dirt can make a radiograph worthless. Industry codes specify that all radiographic darkrooms and equipment must be kept clean and in good condition.
6. Avoid cross contamination of stop bath or fixer solutions into the developer solution.

TANK PROCESSING

In tank processing, as shown in Figure 4.6, the processing solutions and wash water are in tanks deep enough for the film to be submerged. The chemicals in the tank must be stirred and the temperature must be checked. All required equipment should be arranged before turning off the white light. All unnecessary materials should be kept away from the processing area. Test safe lights and arrange them for easy viewing. The door to the darkroom should be locked during processing to avoid unwanted white light.

Before processing, the film is removed from the film holders, or cassettes, used during exposure. This removal is accomplished under darkroom conditions to avoid film fog. The film is grasped by its edges or corners to avoid finger prints, bending, wrinkling or crimping during handling. The film is then placed in a processing hanger (rack) that holds the film firmly by each of its four corners, and is designed to fit the dimensions of the processing tanks. Reels/spools are sometimes used instead of hangers or racks for processing film.

Once placed in the processing hanger or on the spool, the film is ready for processing.

1. The solutions readily reach all portions of both sides of the film. Make sure that the chemicals fully cover the film from the top to the bottom.

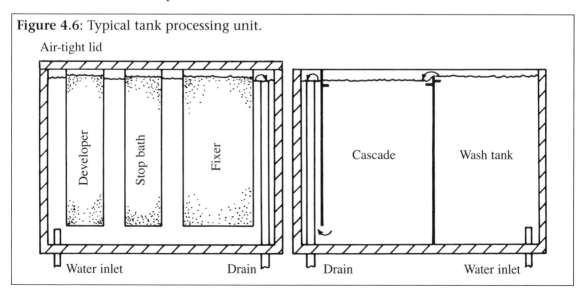

Figure 4.6: Typical tank processing unit.

Air-tight lid

Developer

Stop bath

Fixer

Cascade

Wash tank

Water inlet   Drain   Drain   Water inlet

---

2. Temperature control of the water in which the film is immersed controls the temperature of the solutions. If this feature is not available, then the temperature of each individual tank must be maintained.

Tank Processing Procedures

There are five separate steps in proper tank processing: developing, stop bath, fixing, washing and drying.

Developing

Developing is the chemical process of reducing silver bromide particles in the exposed portion of the film emulsion to metallic silver. This process begins when the film is placed in the developer solution. The amount of silver bromide that is changed is a function of time, the chemical strength and temperature of the solution.

Assuming that the chemical strength remains constant, the density of the radiographic image created through the developing process is proportional to the length of time the film remains in the solution, and directly proportional to the solution temperature (higher temperature equals higher density). To obtain consistent results, the temperature is kept within narrow limits. In practical applications, the contrast and density desired in industrial radiographs is obtained with a solution temperature of 68 °F (20 °C) and a development time between 5 and 8 min. Manufacturers' recommendations and time and temperature charts are consulted whenever doubt exists as to proper developing procedures. For best results, the developing chemicals and the film should be from the same manufacturer.

To avoid fogging, film is never left in the solution beyond recommended time limits. Solution temperature is checked before developing begins because cold solutions result in under development (insufficient film density) caused by retardation of the chemical reaction.

During the development process, waste products of the chemical reaction at the surface of the film are produced, having a higher specific gravity than the solution. These byproducts flow downward and slow down or retard the development of the film areas they pass and cause streaks on the film. For this reason, the film is agitated to obtain uniform development and avoid streaking. When first placed in the solution tank, the film hanger or reel is tapped on the side of the tank to rid the emulsion of air bubbles. Each minute thereafter, until development is completed, the film is shaken vertically and horizontally for a few seconds.

In use, the chemical strength of developer solution grows progressively weaker because of expenditure of the active chemicals in reaction with the silver bromide grains and the buildup of waste reaction products. The rate of this chemical depletion is proportional to the number and density of the films developed. At periodic intervals, determined by the rate of depletion, the activity (development ability) of the solution is tested. If below acceptable standards, the solution is replenished.

Developer activity is tested by processing film radiographed through a stepped wedge. Densities obtained with those of a standard film exposed in the same manner are compared to the film developed in a fresh solution. Similar density comparison results are obtained by cutting the standard film into strips after exposure, each strip containing exposures of all of the steps of the wedge. One strip is then developed in fresh solution and processed as the standard, and the remaining strips are used as solution activity test films. It is good practice to test each day before commencing film processing.

Developers are commercially available in both powder and liquid form. The solution is formed by combining the developer with water. Liquid developer, though more expensive, is much easier to prepare than powder and is normally used. When preparing or replenishing developer solution, the manufacturer's directions should be followed in detail.

Stop Bath

When film is removed from the developing solution, a quantity of the solution remains within the emulsion, and the developing action will continue until the solution is removed. The stop bath, a solution of acetic acid and water, serves to remove this residual developer solution from the film and prevent uneven development and film streaking. The stop bath also neutralizes the alkaline remnants of the developer, permitting the acid in the fixer solution to function in the desired manner.

After development is complete, the film is removed from the developer and allowed to drain for 1 to 2 s. The film is then immersed in the stop bath, which is maintained at the same temperature as the developer solution. The film is agitated in the bath for 30 to 60 s, then removed for transfer to the fixer solution. If no stop bath is available, film is rinsed, after development, in uncontaminated running water for at least 120 s before placing it in the fixer solution.

Stop bath is mixed from commercially available 28% acetic acid or glacial acetic acid; most commonly from the former that is mixed with water, 16 oz (473 mL) of acid to each 1 gal (3.8 L) of solution. When glacial acetic acid is used, the proportions are 4.5 oz (133 mL) of acid to each gallon of solution. Manufacturers' directions must be followed in mixing stop bath, particularly in the handling and preparation of a glacial acetic acid solution. Glacial acetic acid is added slowly to water (never the water to the acid) while stirring constantly.

Stop bath becomes spent after repeated use and is replaced to avoid poor quality radiographs. A fresh stop bath solution is yellow in color and, when viewed under safelight, is almost clear. When indicator stop bath is used, the color changes to a blue-purple color (which appears dark under safelight illumination) indicating the solution needs replacement. About 5 gal (18.9 L) of stop bath will normally treat the equivalent of one hundred 14 by 17 in. (36 by 43 cm) films.

Fixing

If the undeveloped silver bromide remaining in the film emulsion after completion of the developer and stop bath processes is not removed, it will darken on exposure to light and ruin the radiograph. Fixer, a mildly acidic solution, dissolves and removes the silver bromide from the undeveloped portions of the film without affecting the developed portion. It also hardens the emulsion gelatin, permitting warm air drying.

When first placed in the fixer solution, the film becomes clouded as a result of the dissolution of the silver bromide. In time (dependent on the strength of the fixer solution) the film clears, but the dissolution and hardening processes are still going on. The minimum time required for fixing is twice the amount of time necessary to clear the film. It should not exceed 15 min. Longer fixing time, indicative of a weak solution, can cause abnormal swelling of the film emulsion, improper hardening, overly long drying times and loss of lesser film densities. Improper fixing shortens the archival length of the film.

Stop bath and fixer solution should be ±5 degrees of the developer temperature. Particular care is taken to avoid high temperatures that cause the emulsion to wrinkle or slough off. When first placed in the solution, and at 120 s intervals thereafter until fixing is completed, the film is agitated.

Fixer solution becomes depleted through dilution by the stop bath or rinse water carried on the films, and by the accumulation of dissolved silver salts. The solution may be replenished by removing some of the solution and replacing it with undiluted fixer.

There is, however, a limit to the effectiveness of replenishment, and after two or three replenishments the solution is discarded and replaced. The frequency of replenishment and replacement of fixer solution is determined by the acidity of the solution as evidenced by the length of time required for film fixation. It is directly proportional to the number of films processed.

Fixers are commercially available in both powder and liquid form. The fixer solution is formed by combining the fixer with water. Liquid fixer is easier to handle and is most commonly used. In preparing or replenishing fixer solution, the manufacturer's directions should be followed in detail.

Washing

After fixing, films undergo a washing process to remove the fixer from the emulsion. The film is thoroughly immersed in running water when available, so that all of the emulsion surface is in contact with constantly changing water.

Mobile darkrooms are not normally equipped with running water, so daily changing of the rinse water is recommended. The wash tank should be large enough to handle the number of films going through the developing and fixing processes without crowding, and the hourly flow of water should be between four and eight times the volume of the tank.

Each film is washed for a period of time equal to twice the fixing time. When a number of films are proceeding through the processing cycle, each film is first placed in the drain end of the tank, and then progressively moved toward the intake. This procedure ensures that the last wash any film receives is with fresh water.

Hypo Clearing Agent

A hypo clearing agent may be used to help remove the wetting agent or water from the film. The use of such an agent speeds up film washing by improving its efficiency and reduces the amount of water required.

Temperature

The temperature of the water in the wash tank is an important factor of the wash efficiency. Best results are obtained with a water temperature between 65 and 70 °F (18.3 and 21.1 °C), because higher temperatures can cause the same damaging effects as those of high temperatures in the processing solutions. At low temperatures, very little washing action takes place.

Wetting

When film is removed from the wash tank, small drops of water cling to the emulsion. If permitted to remain, these drops will cause water marks or streaks on the finished film. To lessen the possibility of water mark damage, film is immersed in a wetting agent and then drained for 60 to 120 s before drying. Wetting agent is often added to the rinse water. Wetting agents, commercially available in aerosol solutions or concentrated solutions, cause the water to drain evenly from the film and aids in drying time.

Drying

The final step of film processing is drying, usually accomplished by hanging the film in a drying cabinet. Drying cabinets are designed to permit flow of heated and filtered air to reach both sides of the film. If no drying cabinet is available, film may be air dried by hanging in a position where air circulates freely.

Automatic Film Processing

Automatic film processing machines are processing systems built around film, chemicals and mechanics. They are used wherever the volume of work makes them economical. The machines accomplish all required processing, and the only manual operation necessary is loading and unloading the film.

Though the processing steps used in an automatic unit are the same as those for tank processing, the entire processing cycle is completed in less than 15 min. This high speed processing is made possible by special chemicals, continual agitation of the film, maintenance of all solutions at relatively high temperatures and drying with jets of heated air. When properly maintained and operated, automatic film processing units consistently produce radiographs of much higher quality than those processed manually. As with all mechanical parts, cleanliness is of utmost importance.

DARKROOM FACILITIES

Darkroom facilities may consist of a single room where all steps of film handling and processing are accomplished, or of a series of rooms each designed for a specific activity. The location, size and design of the facilities are dependent on the volume and type of work to be done. The location of equipment within the facility is designed to facilitate the logical flow of film through the processing cycle. Two requirements must be satisfied in the construction of a darkroom: it must be lighted with safelight of an intensity sufficient for processing operations without endangering film by exposure to light, and it must be protected against light from outside sources.

Safelights

The placement of safelights in the darkroom is determined by the need for maximum protection against light in the areas where unexposed film is handled (the loading bench). Adequate but less protection is needed in the developing and fixing areas, and normal white light is permitted in washing and drying areas. However, darkroom lighting is usually consistent throughout the area. Safelights of correct wattage (normally not higher than 15 W) properly filtered and at the correct distance from the film can be used in all of these areas.

Safelight installations can be determined safe only through testing. The simplest test for safelights is exposure of film to the light under time and distance conditions equivalent to those encountered during normal film handling. A portion of the test film is protected by opaque material during the exposure. After standard processing, if there is no density difference between the exposed and protected portions of the film, the light is safe.

Protection Against Outside Light

Protection of darkroom spaces against outside light penetrating through entrances is a matter of proper safeguarding through a door locked from the inside, a light lock made with double or revolving doors or a labyrinth entrance. Light tight ventilators are used to prevent light entry through the darkroom ventilating system.

Walls, Ceiling and Floor

The walls and ceiling of the darkroom are usually painted with semigloss paint of a light color that reflects a maximum amount of safelight. The walls in the areas where chemicals may splatter are protected with ceramic tile or glass. Darkroom floors are usually covered with a chemical resistant, waterproof and slipproof material.

Automatic Processing Darkroom

When automatic processing equipment is used, darkroom facilities are designed to accommodate the machine. Because all of the processing takes place within the machine, only the handling of unexposed film and unprocessed film requires darkroom conditions. Usually the machine is installed through a wall so that the loading end is within a darkroom and the remainder in an open area.

DARKROOM EQUIPMENT

The loading bench, film storage cabinets and bins, processing tanks and film driers are standard darkroom equipment. Handling of unprocessed film, loading and unloading of film holders and loading of processing hangers are all accomplished at the loading bench. The storage facilities for holders and hangers and light tight film storage bins are located in the loading bench area. This area, in which all dry activities of film handling take place, should be readily accessible to, but at some distance from, the processing tanks. The dry and wet areas of the darkroom are separated to prevent inadvertent water or chemical damage to film.

Processing Tanks

The processing tanks used in the developing, stop bath, fixing and washing processes are located in the wet area of the darkroom. The tanks are aligned in the order of processing. The relative sizes of the tanks fixes the amount of work that can be done. Developer and stop bath tanks should be the same size, fixer tanks twice as large and wash tanks at least four times as large.

Drying Cabinets

Film drying cabinets should have a filtered air intake, film hanger racks, exhaust fan and a heating element. Because drying is the last processing step, the drier may be conveniently located for ease in film handling.

CLEANLINESS

Cleanliness is of great importance during the entire radiographic process. Film should be handled with care. White, lint free gloves should be used during loading and unloading of film holders, and mounting of film in processing hangers. Film holders, film and screens should only be handled in clean surroundings.

Images of dirt, lead chips, scratched or nicked screens, handling crimps, scratches and nicks on the film may result in a worthless radiograph. Similarly, chemical stains and streaks ruin a radiograph. The film processing area must be kept immaculately clean and access limited to those who work in the area.

Chemical contamination of the area can ruin the film, so it is advisable to store chemicals in a separate area until they are used. Floors must be kept clean at all times. High humidity assists in preventing static electricity and static marks on film. To avoid creating static, do not slide film out of film boxes or cassettes too quickly. Nylon and other fabrics that encourage static electricity should be avoided by the radiographer.

Chapter 5

Safety

INTRODUCTION

This chapter is designed to present some of the basic radiographic safety procedures, protection devices and detection equipment. It is not an interpretation of government regulations, nor is it to be considered as a complete safety guide. The radiographer is cautioned to be aware of the latest effective safety regulations. For the most recent rules and regulations, contact a radiation safety officer, the United States Nuclear Regulatory Commission (NRC) or the regulators in individual states in the United States.

Most of the effects of radiation on the human body are known and predictable. Radiation safety practices are based on these effects and the characteristics of radiation. Because radiation cannot be detected by any of the human senses and its damaging effects do not become obvious immediately, personnel protection is dependent on detection devices and through the proper use of time, distance and shielding.

The NRC enforces safety regulations covering the handling and use of radioisotopes. The Department of Transportation (DOT), the Interstate Commerce Commission (ICC), the Civil Aeronautics Board (CAB) and the United States Coast Guard (USCG) enforce safety regulations covering the transportation of radioactive material.

Some states have similar regulations covering use, handling and transportation of radioactive material. These are called *agreement states*. All of these regulations are designed to limit radiation exposure to safe levels, and to afford protection for the general public. This government emphasis on safety practices indicates the mandatory nature of sure and certain safety practices in all radiation areas. The radiographer, who is employed by a licensee of the NRC, or who is employed by a licensee of an agreement state, must have knowledge of, and comply with, all relevant regulations.

UNITS OF RADIATION DOSE MEASUREMENT

For radiation safety purposes, the combined (cumulative) effect on the human body of radiation exposure is of primary concern. Because the damaging effects of radiation to living cells are dependent on both the type and the energy of the radiation to which they are exposed, it is impractical only to measure radiation quantitatively. For this reason, exposure is first measured in physical terms; then, a factor allowing for the relative biological effectiveness of different types and energies of radiation is applied. The units used to measure radiation exposure are defined as follows.

Roentgen

Exposure to radiation is measured in roentgens (R) or sieverts (Sv). The roentgen (r) is the physical unit measure of the ionization of air by X-radiation or gamma radiation. It is defined as the quantity of radiation that will produce one electrostatic unit (esu) of charge in one cubic centimeter of air at standard pressure and temperature. It is a special unit of exposure. One roentgen of radiation represents the absorption by ionization of about 83 ergs of radiation energy per gram of air. In practical application, the milliroentgen (mR), one thousandth of a roentgen, is often used. One thousand milliroentgens equals one roentgen (1000 mR = 1 R).

Radiation Absorbed Dose

The radiation absorbed dose (rad) is the unit of measurement of radiation absorption by humans. It is considered a special unit of absorbed dose. It represents an absorption of 100 ergs (measurement of work or energy) of energy per gram of irradiated tissue, at the place of exposure. The roentgen applies only to X-rays and gamma rays; the radiation absorbed dose applies to any type of radiation (100 rad = 1 Gray [Gy]).

Quality Factor

The value assigned to various types of radiation, determined by the radiation's effect on the human body, is called *quality factor*. Quality factor values have been calculated by the National Committee on Radiation Protection, as shown in Table 5.1.

Table 5.1: Quality factor values.	
Radiation	**Quality factor**
X-ray	1
Gamma ray	1
Beta particles	1
Thermal neutrons	5
Fast neutrons	10
Alpha particles	20

Roentgen Equivalent Mammal

The roentgen equivalent mammal (rem) is the unit used to define the biological effect of radiation on humans. It represents the absorbed dose, in rads, multiplied by the quality factor of the type of radiation absorbed.

Radiation safety levels are established in terms of roentgen equivalent man dose. The calculating of roentgen equivalent man dose of X-radiation and gamma radiation is simplified by two facts: the roentgen dose is equivalent to the radiation absorbed dose; and the quality factor of both X-radiation and gamma radiation is one. A measurement of roentgen dose thus becomes a measurement of roentgen equivalent mammal dose (1 rad = 1 rem).

INTERNATIONAL SYSTEM OF UNITS (SI) MEASUREMENTS

Because of existing practice in certain fields and countries, the International Committee for Weights and Measures (CIPM, Comité Internationale des Poids et Mesures) permitted the units curie, roentgen, rad and rem to continue to be used with the SI measurement system until 1998. The National Institute of Standards and Technology (NIST) strongly discourages the continued use of curie, roentgen, rad and rem. The American National Standards Institute (ANSI), the American Society for Testing and Materials (ASTM), the Institute of Electrical and Electronics Engineers (IEEE), the International Organization for Standardization (ISO) and the American Society for Nondestructive Testing (ASNT) all support the replacement of older English units with SI units. However, NRC and State Regulations are still stated in the units: curies, roentgen and rem and are the units most used by radiographers in the United States.

Becquerel Replaces Curie

The original unit for radioactivity was the curie (Ci), the radiation emitted by one gram of radium. Eventually all equivalent radiation from any source was measured with this same unit. It is now known that a curie is equivalent to 3.7×10^{10} disintegrations per second. In SI, the unit for radioactivity is the becquerel (Bq), which is one disintegration per second. Because billions of disintegrations are required in a useful source, the multiplier prefix giga (10^9) is used and the unit is normally seen as gigabecquerel (GBq). One curie is equal to 37 GBq.

Coulomb per Kilogram Replaces Roentgen

The unit for quantity of electric charge is the coulomb (C), where 1 C = 1 ampere $\times$ 1 second. The original roentgen (R) was the quantity of radiation that would ionize 1 cm^3 of air to 1 electrostatic unit of electric charge, of either sign. It is now known that a roentgen is equivalent to 258 microcoulombs per kilogram of air

$(258 \ \mu C \cdot kg^{-1}$ of air). This corresponds to 1.61×10^{15} ion pairs per 1 kg of air, which has then absorbed 8.8 mJ (joule) (0.88 rad, where rad is the obsolete unit for radiation absorbed dose, not the SI symbol for radian).

Gray Replaces Rad

The roentgen (R) was an intensity unit but was not representative of the dose absorbed by human tissue in a radiation field. The radiation absorbed dose (rad) was first created to measure this quantity and was based on the erg, the energy unit from the old centimeter-gram-second (CGS) system. In the SI system, the unit for radiation dose is the gray (Gy), and 1 Gy = 100 rad. The gray is useful because it applies to doses absorbed by human tissue at a particular location. It is expressed in energy units per mass of matter or in joules per kilogram ($J \cdot kg^{-1}$). The mass is that of the absorbing human tissue.

Sievert Replaces Rem

The SI system's unit for the dose absorbed by the human body (formerly rem for roentgen equivalent man; also known as *ambient dose equivalent*, *directional dose equivalent*, *dose equivalent*, *equivalent dose* and *personal dose equivalent*) is similar to the gray but includes quality factors dependent on the type of radiation. This absorbed dose has been given the name sievert (Sv) but its dimensions are the same as the gray, that is, 1 Sv = 1 $J \cdot kg^{-1}$ (1 Sv = 100 rem).

Maximum Permissible Dose

It is impossible and impractical to safeguard radiography personnel from some exposure to radiation. Permissible dose is defined by the NIST as the dose of ionizing radiation that, in the light of present knowledge, is not expected to cause appreciable bodily injury to a person at any time during the lifetime.

Occupational annual dose limits as established by the NRC for classified workers is the more limiting of the following.

1. Total effective dose equivalent being equal to 5 rem (0.05 Sv).

Or

2. The sum of the deep dose equivalent and the committed dose equivalent to any individual organ or tissue other than the lens of the eye being equal to 50 rem (0.5 Sv);
3. The lens dose equivalent of 15 rem (0.15 Sv); and
4. A shallow dose equivalent of 50 rem (0.5 Sv) to the skin of the whole body or to the skin of any extremity.

Maximum radiation dose in any period of one calendar year to an individual in a restricted area is normally limited to 5 rem (0.05 Sv).

Occupational annual dose limits for minors are 10% of the annual dose limits for adult workers.

Any woman voluntarily declaring pregnancy must do so in writing, and it remains in effect until she is no longer pregnant. The licensee shall ensure that the dose equivalent to the embryo/fetus does not exceed 0.5 rem (5 mSv) during the entire pregnancy by monitoring the lower torso region.

Dose limits to the general public shall not exceed 0.002 rem or 2 mrem (0.02 mSv) in any 1 h or exceed 0.5 rem or 500 mrem (5 mSv) in one year.

PROTECTION AGAINST RADIATION

Three cardinal principles govern safety practices for controlling body exposure to radiation: time, distance and shielding. Safe radiographic techniques and radiographic installations are designed by applying these principles.

Allowable Working Time
The amount of radiation absorbed by the human body is directly proportional to the time the body is exposed. A person receiving 2 mrem (0.2 mSv) in 60 s at a given point in a radiation field would receive 10 mrem (1 mSv) in 5 min. Allowable working time is calculated by measuring radiation intensity and substituting it in the following equation.

Eq. 5.1 $$\text{Allowable working time in h/wk} = \frac{\text{permissible exposure in Ci/wk}}{\text{exposure rate in Ci/h}}$$

Working Distance
The greater the distance from a radiation source, the lower the radiation intensity. The inverse square law is used to calculate radiation intensities at various distances from a source. The inverse square law (illustrated in Figure 2.8) is expressed as:

Eq. 5.2 $$\frac{I_1}{I_2} = \frac{D_2^2}{D_1^2}$$

where I_1 and I_2 are intensities at distances D_1 and D_2, respectively.

The following examples illustrate methods used to calculate radiation intensities in terms of dose rate. Table 5.2 lists dose rates of commonly used radioisotopes referred to in the examples.

Table 5.2: Radioisotope dose rate.	
Radioisotope	**Dose rate R/h/Ci (Sv/h/Ci) at 1 ft (0.3 m) emissivity**
Co-60	14.0 (0.14)
Ir-192	5.2 (0.052)
Cs-137	3.4 (0.034)
Tm-170	0.027 (0.00027)

Example 1: Given a 12 Ci Cs-137 source, what is the emission at 3 ft?

Step 1: From Table 5.2, the dose rate of Cs-137 is 3.4 R/h/Ci at 1 ft. Thus, the dose rate of a 12 Ci source at 1 ft is 12×3.4 or 40.8 R/h.

Step 2: $I_2 = 40.8$ R/h
$D_2 = 1$ ft
$D_1 = 3$ ft

Step 3: Substituting in the inverse square law equation:

$$\frac{I_1}{40.8} = \frac{1^2}{3^2}$$

Step 4: Solving for I_1:

$$I_1 = 40.8 \times \frac{1^2}{3^2} = 40.8 \times \frac{1}{9} = 4.53\, R/h$$

Example 2: A 35 Ci source of Ir-192 is used at distance of 20 ft from a radiographer. What dose rate will the radiographer be exposed to?

Step 1: From Table 5.2, the dose rate of Ir-192 is 5.2 R/h/Ci at 1 ft; thus, the dose rate of a 35 Ci source at 1 ft is 35×5.2 or 182 R/h.

Step 2: $I_2 = 182$ R/h
$D_2 = 1$ ft
$D_1 = 20$ ft

Step 3: Substituting in the inverse square law equation:

$$\frac{I_1}{182} = \frac{1^2}{20^2}$$

Step 4: Solving for I_1:

$$I_1 = 182 \times \frac{1^2}{20^2} = 182 \times \frac{1}{400} = 0.455\ R/h$$

Example 3: In Example 2, at what distance from the source would the dose rate to the radiographer be to receive only 3 mR/h?

Step 1: $I_2 = 182$ R/h, or 182 000 mR/h
$I_1 = 3$ mR/h
$D_2 = 1$ ft

Step 2: Substituting in the inverse square law equation:

$$\frac{3}{182\,000} = \frac{1^2}{D_1^2}$$

Step 3: Solving for D_1:

$$D_1^2 = \frac{182\,000 \times 1^2}{3} = 60\,666$$

$$D_1 = \sqrt{60\,666} = 246 + feet$$

Tables, such as Table 5.3 that list the dose rates per Ci at various distances from a source, are derived by application of the inverse square law.

Table 5.3: Radioisotope dose rates per curie versus distance. Dose rate in R/h/Ci.

Distance	Co-60	Ir-192	Cs-137	Tm-170	Se-75
1 ft	14.0	5.2	3.4	0.027	2.2
2 ft	3.5	1.3	0.85	0.068	0.53
4 ft	0.875	0.33	0.21	0.016	0.1375
8 ft	0.219	0.081	0.053	0.004	0.0343
10 ft	0.14	0.052	0.034	0.0027	0.022

Example 4: Given the emission of Co-60 as 14 R/h/Ci at 1 ft, what is the dose rate per Ci at 2, 4, 8, etc. ft?

Step 1: $I_2 = 14$ R/h/Ci
$D_2 = 1$ ft
$D_1 = 2$ ft

Step 2: Substituting in the inverse square law equation:

$$\frac{I_1}{14} = \frac{1^2}{2^2}$$

Step 3: Solving for I_1:

$$I_1 = 14 \times \frac{1^2}{2^2} = 14 \times \frac{1}{4} = 3.5\,R/h/Ci$$

Step 4: Solve for dose rates per Ci at other distances in similar fashion.

All of the foregoing examples are based on gamma radiation; however, the same principles for calculating dose rate or radiation intensity hold for X-radiation. In determining X-radiation intensities, it is necessary to measure intensity at a known distance with predetermined amperage and kilovoltage settings, and then apply the inverse square law.

Any change in X-ray machine settings requires a new intensity measurement and recalculation. Intensity and dose rate calculations based on the inverse square law should never be accepted as exact. Radiation intensity at any point is the sum of the primary radiation and the secondary (scatter) radiation at that point. Only under ideal conditions of no scatter and in a complete vacuum are calculated intensities exact.

Shielding

Lead, steel, water and concrete are materials commonly used as shielding to reduce personnel exposure. Because all of the energy of X-radiation or gamma radiation cannot be stopped by shielding, it is practical to measure shielding efficiency in terms of half value layers.

The half value layer is that amount of shielding that will stop half of the radiation of a given intensity. Similarly, shielding efficiency is often measured in tenth value layers. A tenth value layer is that

Table 5.4: Approximate X-ray half value layers.

Shielding material	Half value layer for X-ray tube potential of:							
	50 kVp	70 kVp	100 kVp	125 kVp	150 kVp	200 kVp	250 kVp	300 kVp
Lead (mm)	0.05	0.15	0.24	0.27	0.29	0.48	0.9	1.4
Concrete (in.)	0.168	0.33	0.59	0.79	0.88	1	1.1	1.23

Table 5.5: Approximate gamma ray half and tenth value layers.

Shielding material, inch (centimeter)	Radioisotope source					
	Co-60		Ir-192		Cs-137	
	1/10	1/2	1/10	1/2	1/10	1/2
Lead	1.62 (4.1)	0.49 (1.2)	0.64 (1.6)	0.19 (0.5)	0.84 (2.1)	0.25 (0.6)
Steel	2.9 (7.4)	0.87 (2.2)	2 (2.5)	0.61 (1.5)	2.25 (5.7)	0.68 (1.7)
Concrete	8.6 (21.8)	2.6 (6.6)	6.2 (15.7)	1.9 (4.8)	7.1 (18)	2.1 (5.3)
Aluminum	8.6 (21.8)	2.6 (6.6)	6.2 (15.7)	1.9 (4.8)	7.1 (18)	2.1 (5.3)

amount of shielding that will stop nine tenths of the radiation of a given intensity, as shown in Tables 5.4 and 5.5.

Half and tenth value layers are, in all cases, determined by experiment and actual measurement. The radiographer should rely only on actual measurement to determine the effectiveness of any shielding.

The following examples illustrate the application of half value layer information.

Example 5: A 200 kVp X-ray machine must be located so that the primary radiation is directed toward an adjacent occupied room. Without shielding, the dose rate in the adjacent room is 500 times the acceptable safe limit. How thick of a concrete wall is required to reduce the dose rate in the adjacent room to a safe value?

> **Step 1**: Because one half value layer reduces dose rate by a factor of 1/2; two half value layers by $1/2 \times 1/2$ or 1/4; three half value layers by $1/2 \times 1/2 \times 1/2$ or 1/8; etc., then 9 half value layers will reduce dose rate by a factor of 1/512, and 9 half value layers will reduce the dose rate to an acceptable safe limit.

> **Step 2**: From Table 5.4, the concrete half value layer for 200 kVp radiation is 1 in. Thus, 9 in. (22.9 cm) of concrete shielding is required to reduce the dose rate to an acceptable safe value.

Example 6: In a previous example, it was found that the dose rate for a technician 20 ft (6.1 m) from a 35 Ci Ir-192 source was 516.25 mR/h. If the technician must remain at the same location, how much lead shielding is required to reduce the dose rate to 3 mR/hr?

> **Step 1**: The desired dose rate is 3 mR/h; therefore the original dose rate of 516.25 must be reduced 516.25/3 or 172 times.

> **Step 2**: Seven half value layers reduces dose rate by a factor of $1/2^7$ or 128 times, but 8 half value layers reduces it by a factor of $1/2^8$ or 256 times. Thus, 7 half value layers will not provide the required shielding, but 8 half value layers will.

> **Step 3**: From Table 5.5, the lead half value layer for Ir-192 radiation is 0.19 in. Therefore, 8×0.19 or 1.52 in. of lead shielding is required to reduce the dose rate to an acceptable safe value.

Exposure Area

Enclosed exposure areas should consist of a room with thick concrete walls, completely lined with lead, or other suitable shielding material of sufficient thickness for protection. If the construction of such a room is not feasible, then the equipment should be housed in a suitably shielded cabinet large enough to also house the test objects.

Controls should be located outside the exposure area. To reduce the possibility of excessive radiation in occupied spaces, the exposure area must be isolated. If neither a room nor a cabinet is available, a combination of shielding that safely encloses the radiation equipment, test object and the film is required.

It is not always practical to bring the test object to the shielded exposure area. When radiography must be accomplished under these conditions (e.g. temporary job sites), the three safety factors (time, distance and shielding) must be taken into account. Safe distances, in relation to exposure, must be determined and adequately marked with guard rails or ropes placed to enclose the radiation area, the area must be clearly marked with legible and noticeable radiation warning signs, and only monitored radiographers are permitted in the radiation area.. Sufficient shielding, if required, must be placed to protect the radiographer and others who must remain in the vicinity.

When radiography is practiced outside a designated shielded exposure area, the simplest, most effective safety consideration is distance. All personnel must be kept at a safe distance from the radiation source.

X-Ray Tube Shielding

In theory, the lead housing around an X-ray tube effectively shields, to safe levels, all primary radiation except the useful beam. Practically, this is not always the case, and the only way to ensure the safeness of an X-ray tube is to measure leakage (unwanted) radiation. To limit the unwanted radiation, the area of primary radiation should be fixed by a cone or diaphragm at the tubehead.

Radiation Protective Construction

The most common materials used to protect against radiation are lead and concrete. Shielding measurements are usually expressed in terms of thickness. Particular care must be exercised to ensure leak proof shielding. Adjacent sheets of lead must be overlapped, and nails or screws that pass through the lead must also be covered with lead. Pipes, conduits and air ducts passing through the walls of the shielded area must be completely shielded. Figure 5.1 illustrates good lead shielding construction practices.

The thickness of the lead shield is dependent on the energy of the radiation requiring shielding and the use (occupancy) of the surrounding areas. If the spaces above, below and around the exposure area are occupied, then all of the exposure area — wall, ceiling and floor — must be shielded. If the room is on the top floor of the building, it is not necessary to shield all of the ceiling. Similarly, if the room is on the bottom floor, not all of the floor needs shielding. The methods of partial floor shielding shown in Figure 5.1 also apply to partial shielding of a ceiling. In either case,

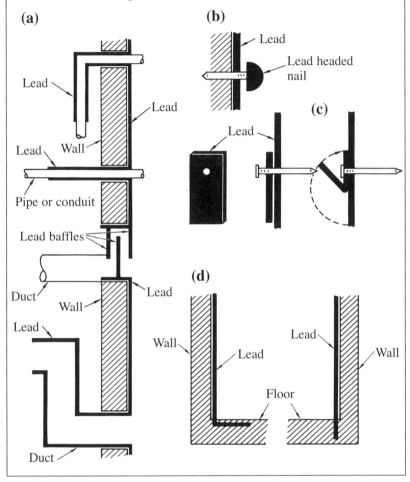

Figure 5.1: Radiation protection constructions: (a) methods of shielding when pipes, ducts or conduits must pass through walls of an X-ray room; (b) method of sealing nail or screw holes in lead protection for a lead headed nail; (c) method of sealing nail or screw holes in lead protection for a lead strip folded over nail heads; and (d) methods of handling protection at floor level when protection is not needed over entire floor.

(a)

Lead

Lead

Lead Wall

Pipe or conduit

Lead baffles

Duct

Wall

Lead

Lead

Duct

(b)

Lead

Lead headed nail

(c)

Lead

(d)

Wall

Lead

Lead

Wall

Floor

the partial shielding prevents radiation escaping above or below the wall and scattering into an adjacent area.

Though lead is the most efficient of the easily available shielding materials, other structural materials such as concrete and brick are often used. At voltages greater than 400 kV, the thickness of lead shielding would be so great as to make it difficult to fasten the lead to the walls, as well as cost prohibitive. At these higher voltage potentials, concrete is used as shielding because of its relative effectiveness and its construction simplicity.

Gamma Ray Requirements

Special gamma radiation protection requirements are based on two factors: gamma radiation is very penetrating; and the required protective shielding is excessively thick and heavy, as shown in Table 5.5. Gamma radiation cannot be shut off, and protection from gamma sources must be provided at all times.

A combination of distance and shielding is usually used during gamma radiography. The radiation area is roped off and clearly marked with legible and noticeable radiation warning signs, and only monitored radiographers are permitted in the radiation area. The extent of the danger zone is based on calculations of safe distance as determined by the source activity using the inverse square law formula. In calculating the area of the danger zone, the possible effects of scatter radiation are considered, and the calculations are confirmed by intensity measurements.

The continuous gamma radiation from radioisotopes necessitates strict accountability of radioactive sources. When not in use, they are stored in conspicuously labeled storage containers. After every use, readings with survey meters are taken to ensure that the source is safely stored, and the projector unit is not emitting excessive radiation. Projectors contain shielding (called *pigs*) typically made out of lead or depleted uranium for shielding purposes.

UNITED STATES NUCLEAR REGULATORY COMMISSION

The previously discussed safety precautions are nonspecific in nature. Handling, storage and use of radioisotopes are regulated by the NRC. The regulations are published in the *Code of Federal Regulations, Title 10, Chapter I*, parts 19, 20 and 34. These three parts of the code are also published in the *United States Nuclear Regulatory Commission Licensing Guide*. The following regulations are subject to change and are presented here only to familiarize the reader.

The NRC also provides *NRC Form-4: Cumulative Occupational Dose History*, and *NRC Form-5: Occupational Dose Record for a Monitoring Period*. These forms are shown in Figures 5.2 and 5.3. They should be used to monitor all radiation exposure.

Occupational Radiation Exposure Limits

Limitations on individual dosage are specified in Table 5.6. Doses greater than those specified in Table 5.6 may be permitted if the following factors apply.

1. During any calendar year, the dose to the whole body does not exceed 5 rem (0.05 Sv).
2. The individual's accumulated occupational dose has been recorded on *NRC Form-4: Cumulative Occupational Dose History* and the concerned individual has signed the form.

Figure 5.2: *NRC Form-4: Cumulative Occupational Dose History.*

NRC FORM 4
(9-2004)
10 CFR PART 20

U.S. NUCLEAR REGULATORY COMMISSION

CUMULATIVE OCCUPATIONAL DOSE HISTORY

APPROVED BY OMB NO.3150-0005   EXPIRES: 09/30/2007

Estimated burden per response to comply with this mandatory collection request: 30 minutes. This information is required to record an individual's lifetime occupational exposure to radiation to ensure that the cumulative exposure to radiation does not exceed regulatory limits. Send comments regarding burden estimate to the Records and FOIA/Privacy Services Branch (T-5 F52), U.S. Nuclear Regulatory Commission, Washington, DC 20555-0001, or by internet e-mail to infocollects@nrc.gov, and to the Desk Officer, Office of Information and Regulatory Affairs, NEOB-10202, (3150-0005), Office of Management and Budget, Washington, DC 20503. If a means used to impose an information collection does not display a currently valid OMB control number, the NRC may not conduct or sponsor, and a person is not required to respond to, the information collection.

1. NAME (LAST, FIRST, MIDDLE INITIAL)

2. IDENTIFICATION NUMBER

3. ID TYPE

4. SEX
☐ MALE
☐ FEMALE

5. DATE OF BIRTH (MM/DD/YYYY)

| 6. MONITORING PERIOD (MM/DD/YYYY - MM/DD/YYYY) | 7. LICENSEE NAME | 8. LICENSE NUMBER | 9. ☐ RECORD ☐ ESTIMATE ☐ NO RECORD | 10. ☐ ROUTINE ☐ PSE |
| 11. DDE | 12. LDE | 13. SDE, WB | 14. SDE, ME | 15. CEDE | 16. CDE | 17. TEDE | 18. TODE |

| 6. MONITORING PERIOD (MM/DD/YYYY - MM/DD/YYYY) | 7. LICENSEE NAME | 8. LICENSE NUMBER | 9. ☐ RECORD ☐ ESTIMATE ☐ NO RECORD | 10. ☐ ROUTINE ☐ PSE |
| 11. DDE | 12. LDE | 13. SDE, WB | 14. SDE, ME | 15. CEDE | 16. CDE | 17. TEDE | 18. TODE |

| 6. MONITORING PERIOD (MM/DD/YYYY - MM/DD/YYYY) | 7. LICENSEE NAME | 8. LICENSE NUMBER | 9. ☐ RECORD ☐ ESTIMATE ☐ NO RECORD | 10. ☐ ROUTINE ☐ PSE |
| 11. DDE | 12. LDE | 13. SDE, WB | 14. SDE, ME | 15. CEDE | 16. CDE | 17. TEDE | 18. TODE |

| 6. MONITORING PERIOD (MM/DD/YYYY - MM/DD/YYYY) | 7. LICENSEE NAME | 8. LICENSE NUMBER | 9. ☐ RECORD ☐ ESTIMATE ☐ NO RECORD | 10. ☐ ROUTINE ☐ PSE |
| 11. DDE | 12. LDE | 13. SDE, WB | 14. SDE, ME | 15. CEDE | 16. CDE | 17. TEDE | 18. TODE |

| 6. MONITORING PERIOD (MM/DD/YYYY - MM/DD/YYYY) | 7. LICENSEE NAME | 8. LICENSE NUMBER | 9. ☐ RECORD ☐ ESTIMATE ☐ NO RECORD | 10. ☐ ROUTINE ☐ PSE |
| 11. DDE | 12. LDE | 13. SDE, WB | 14. SDE, ME | 15. CEDE | 16. CDE | 17. TEDE | 18. TODE |

| 6. MONITORING PERIOD (MM/DD/YYYY - MM/DD/YYYY) | 7. LICENSEE NAME | 8. LICENSE NUMBER | 9. ☐ RECORD ☐ ESTIMATE ☐ NO RECORD | 10. ☐ ROUTINE ☐ PSE |
| 11. DDE | 12. LDE | 13. SDE, WB | 14. SDE, ME | 15. CEDE | 16. CDE | 17. TEDE | 18. TODE |

19. SIGNATURE OF MONITORED INDIVIDUAL

20. DATE SIGNED

21. CERTIFYING ORGANIZATION

22. SIGNATURE OF DESIGNEE

23. DATE SIGNED

NRC FORM 4 (9-2004)

PRINTED ON RECYCLED PAPER

Figure 5.3: *NRC Form-5: Occupational Dose Record for a Monitoring Period.*

NRC Form-5: Occupational Dose Record for a Monitoring Period is shown in Figure 5.3. This form must be completed annually and is the source of the information recorded on *NRC Form-4: Cumulative Occupational Dose History.*

Table 5.6: Maximum permissible dose.

Annual exposure in rem (sievert)	
Total effective dose equivalent	5 (0.05)
Deep dose equivalent and committed dose equivalent to any individual organ or tissue other than the lens of the eye	50 (0.5)
Lens of the eye dose equivalent	15 (0.15)
Shallow dose equivalent to the skin and extremities	50 (0.5)

Levels of Radiation in Unrestricted Areas

Exposure limits in unrestricted areas are listed in Table 5.7. These exposure limits are based on an individual being continually present in the area, and thus represent maximum radiation levels permitted.

Table 5.7: Exposure limits in unrestricted areas.

Exposure time	Exposure limit millirem (millisievert)
1 hour	2 (0.02)
1 calendar year	100 (1)

Personnel Monitoring

During radiographic operations, radiographers and their assistants shall wear personnel dosimeters, film badges, thermoluminescent dosimeter (TLDs) or optically stimulated luminescence (OSL) badges, and direct reading dosimeters, pocket dosimeters or electronic personal dosimeters. Pocket dosimeters and electronic personal dosimeters shall be capable of measuring exposures from 0 to 200 mR (0 to 2mSv). They should be read periodically through the day during radiographic operations, and indicated doses should be recorded. If a pocket dosimeter is discharged beyond its range, or the electronic personal dosimeter reads greater than 200 mR (2 mSv), the personnel dosimeter of the individual shall be processed immediately.

Caution Signs, Labels and Signals

The radiation symbol is shown in Figure 5.4. Signs bearing this symbol must be placed in conspicuous places in all exposure areas, and on all containers in which radioactive materials are transported, stored or used. On each sign, the word *caution* or *danger* must appear. Other wording required is determined by specific sign use. Area signs bear the phrases *radiation area*, *high radiation area*, *very high radiation area* or *airborne radioactivity area*, as appropriate.

Containers of radioactive materials and areas housing such containers must be marked with signs or labels bearing the radiation symbol and the words *radioactive material*. Special tags bearing the radiation symbol and the phrase, *Danger radioactive material — do not handle*. Company information and a 24 h phone number must be attached the exposure device.

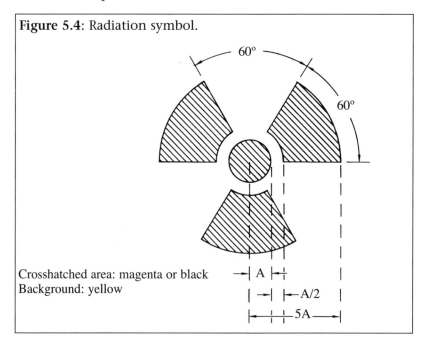

Figure 5.4: Radiation symbol.

Crosshatched area: magenta or black
Background: yellow

Exposure Devices and Storage Containers

Specific regulations provide standards for isotope cameras and exposure devices. Protective standards designed to protect personnel from sealed sources when they are in the off (shielded) position are as follows.

1. Exposure devices must have the name of the company or lab and the location of the office placed in a noticeable site on the device.
2. All of these labels, signs, etc., shall be legible.

Radiation Survey Instrumentation Requirements

For radiographers, it is required that calibrated and operable radiation survey instruments (meters) be available. Unless the operating and emergency procedures exceed the minimum requirements, each exposure device shall be accompanied by one survey meter. If the company's operating and emergency procedures stipulate two or more operable and calibrated survey meters per exposure device, then the more stringent rule is enforced.

The meters shall have a range such that 2 mR (0.02 mSv) per hour through 1 R (0.1 Sv) per hour can be measured.

Radiation Surveys

Specific regulations for required surveys are as follows.

1.   No radiographic operation shall be conducted unless calibrated and operable radiation survey instrumentation is available and used at each site where radiographic exposures are made.
2.   A physical radiation survey shall be made after each radiographic exposure during operation to determine that the sealed source has been returned to its shielded condition. This is known as a *360° sweep*, and the circumference of the exposure device (camera), if it's a crank out device, includes a source tube and the collimator.
3.   A physical radiation survey shall be made to determine that each sealed source is in its shielded position before storing the radiographic exposure device and storage container. These readings shall be recorded, usually on a radiation report survey.

DETECTION AND MEASUREMENT INSTRUMENTS

Various techniques, based on the characteristic effects of radiant energy on matter, are used in detection and measurement devices. Chemical and photographic detection methods are used, as well as methods that measure the excitation effect of radiation on certain materials.

In radiography, however, the instruments most commonly used for radiation detection and measurement rely on the ionization produced in a gas. Because the hazard of radiation is calculated in terms of total dose and dose rate, the instruments used for detection and measurement logically fall into two categories: instruments that measure total dose exposure such as pocket dosimeters, electronic personnel dosimeters, thermoluminescent dosimeters (TLDs) and optically stimulated luminescence (OSL) badges, and instruments that measure dose rate (radiation intensity) such as ionization chambers and geiger-mueller counters. These instruments are known as *survey meters*.

Pocket Dosimeters

The pocket dosimeter, shown in Figure 5.5, is a small device, about the size of a fountain pen. Its operation is based on two principles: like or similar electrical charges repel each other; and radiation causes ionization in a gas.

The essential components of the dosimeter are the metal cylinder, the metal coated quartz fiber electrode consisting of a fixed section and a movable section, the transparent scale, and the lens. The electrode and the cylinder form an electroscope. When a potential voltage (from an external source) is applied between the electrode and the cylinder, the electrode gains a positive charge and the cylinder a negative charge. Simultaneously, the movable portion of the electrode moves away from the fixed portion because they are mutually repellent, each carrying a positive charge. The transparent scale and the lens are so placed that, when the scale is viewed through the lens, the movable portion of the electrode appears as the indicator on the scale. When the dosimeter is properly charged, the indicator will be at zero scale and the dosimeter is ready for use.

When a dosimeter is placed in an area of radiation, ionization takes place in the cylinder chamber. Negative ions are attracted to the electrode, and positive ions are attracted to the cylinder. As the positive charge on the electrode becomes neutralized, the repellent force between the fixed and movable portions decreases. The movable portion moves toward the fixed portion in an amount proportional to the ionization action.

Because the quantity of ionization is determined by the quantity of radiation, the displacement of the movable portion of the electrode is a direct measure of radiation. Pocket dosimeters are designed with a sensitivity that permits them to be scaled in doses from 0 to 200 mR (0 to 2 mSv). Per NRC regulations, they must be calibrated annually.

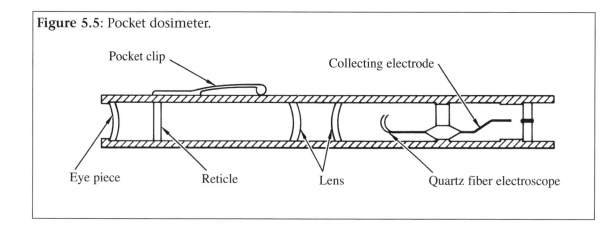

Figure 5.5: Pocket dosimeter.

Pocket clip

Collecting electrode

Eye piece

Reticle

Lens

Quartz fiber electroscope

Personal Electronic Dosimeters

Personal electronic dosimeters, also referred to as *electron dosimeters*, are easy to use, sensitive and have numerous dosimeter functions that can be enabled or disabled. The electronic dosimeter provides dose, dose rate and set point check, and usually operates with one AA battery.

The set points can be preset to definitive alarm points and chirp increments, depending on requirements. For instance, one type ranges from 6 mR to 999R, and the chirp rate is about one chirp per 6 mR of dose received.

The pocket sized monitors provide three-digit digital display of accumulated dose and audible chirp indication of exposure rate. The energy response of the pocked sized monitor for gamma and X-ray is 40 keV to 1.2 MeV. Per NRC regulations, they should be calibrated annually.

Film Badges and Thermoluminescent Dosimeters

The film badge, shown in Figure 5.6, consists of a small film holder equipped with thin lead or cadmium filters, in which a special X-ray film is inserted. The badge is designed to be worn by an individual when in radiation areas and is not to be otherwise exposed.

The use of thermoluminescent dosimeters (TLDs) is another common method of dosimetry, much like the film badge in its outer appearance. It contains a special crystal of lithium fluoride rather than a sheet of film. This crystal has the unique ability to absorb and store energy resulting from interactions with X-rays. When wearing a TLD, incoming photons interact with the crystal, depositing energy into it. The energy is stored within the crystal. The TLD is sent to a laboratory where the crystals are processed. An accurate measurement of dose can be extracted by the amount of energy stored within the crystal. Thermoluminescent dosimeters are not as sensitive to heat, moisture and rough handling as film badges are, but they are more expensive to purchase and process.

After a period of time, usually one month, the film is removed and developed by standard techniques. The density of the processed film is proportional to the radiation received. By use of a densitometer, the density of the film is compared to a set of control films. Through this comparison, an estimate of the amount of radiation received by the individual wearing the badge is made. Film badges and dosimeters each record total radiation received and serve as a check on each other.

Optically Stimulated Luminescence Badges

Optically stimulated luminescence (OSL) badges measure beta, gamma and X-radiation exposure. The OSL is a thin strip of specially formulated aluminum oxide crystalline material. It detects energies from 5 keV to in excess of 40 MeV for photons, 150 keV to in excess of 10 MeV for beta particles and 40 keV to in excess of

Figure 5.6: Typical film badges.

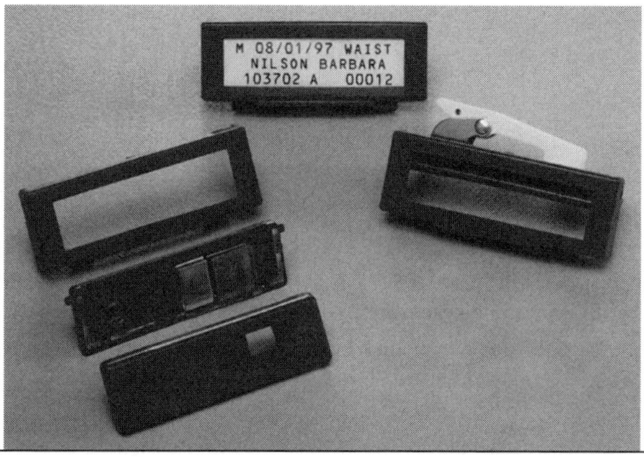

35 MeV for neutrons. The dose measurements range from 1 mrem to 1000 rem.

Ionization Chamber Instruments

Ionization chamber instruments consist of an ionization chamber containing two electrodes, a power supply, usually a battery, which is connected across the electrodes, and an ammeter connected in series with the power supply. When the instrument is exposed to radiation, ionization takes place in the chamber. Individual ions are attracted to the electrode of opposite potential; and, on reaching the electrode, become neutral by removing a charge from the battery.

The flow of current from the battery required to neutralize the ions is measured by the meter that is calibrated in terms of milliroentgen per hour (mR/hr) or millisievert per hour (mSv/hr). The meter may be calibrated in radiographic terms because the flow of current is proportional to the ionization caused by the radiation. In this manner, radiation intensity (dosage rate) is measured. Ionization chamber instruments typically attain an accuracy of ±15% except in low intensity radiation areas. In areas of low intensity radiation, sufficient ionization current is not generated to indicate accurately on the meter. Radiation intensity measurements in areas of low intensity radiation are usually made with geiger-mueller counters. Per NRC regulations, they should be calibrated annually.

Geiger-Mueller Counters

Geiger-mueller counters use a geiger-mueller tube instead of an ionization chamber in a high sensitivity radiation detecting device. The voltage difference between the tube anode and cathode and the gas within the tube creates an environment wherein any ionizing event is multiplied into many such events.

The secondary ionizations are caused by the action of the electrons produced in the first ionization event. This phenomenon of a single ionization producing many in a fraction of a millisecond is known as *gas multiplication*. The resultant amplified pulse of electrical energy is used to cause an audible indication, deflect a meter or light a lamp. Per NRC regulations, geiger-mueller counters should be calibrated annually.

Geiger-mueller counters are typically accurate to ±20% for the quality of radiation to which they are calibrated. In areas of high radiation intensity, geiger-mueller counters have a tendency to saturate and the meter will indicate a false zero reading. For this reason, in areas of suspected high intensity radiation, ionization chamber instruments should be used. Note that this is typically unlikely unless a field of about 1000 R/hr is encountered.

Area Alarm Systems

Area alarm systems consist of one or more sensing elements, usually ionization chambers, whose output is fed to a central alarm meter. The meter is preset so that an audible alarm is sounded, and a visual indication is given (lighted lamp) when permissible radiation levels are exceeded. Area alarm systems are required for shielded room radiography.

ELECTRICAL SAFETY

The radiographer must comply with safe electrical procedures when working with X-ray equipment. Modern X-ray machines use high voltage circuits. Permanently installed X-ray facilities are designed so that personnel trained in safe practices will encounter little electrical hazard; however, portable X-ray equipment requires certain electrical precautions.

Whenever X-ray equipment is being operated or serviced, the following precautions, applicable to both permanent and portable installations, should be observed.

1. Do not turn the power on until setup for exposure is completed.
2. Ensure that grounding instructions are complied with.
3. Regularly check power cables for signs of wear. Replace when necessary.
4. Avoid handling power cables when the power is on.
5. If power cables must be handled with the power on, use safety equipment such as rubber gloves, rubber mats and insulated high voltage sticks.
6. Ensure that condensers are completely discharged before checking any electrical circuit.

RT

LEVEL II

Chapter 6

Specialized Radiographic Applications

INTRODUCTION

A quality radiograph will have low distortion, high definition, high contrast and adequate density where exposure is controlled.

This chapter presents information obtained in the field and laboratory on different exposure techniques. This information allows radiographers to maintain reasonable control over the different radiographic tasks. The radiographer, with a basic knowledge and understanding of the radiographic process and the ability to use the data available, can devise effective procedures for the radiography of different test objects.

Proper film processing, as described in Chapter 4, is an essential aspect of proper radiographic practice. An ideal exposure technique can result in a worthless radiograph if the film is improperly processed.

SELECTION OF EQUIPMENT

Selection of equipment for a particular test consists of the following related decisions.

1.   Selection of radiography as a test method.
2.   Selection of X-radiography or gamma radiography.
3.   Selection of specific X-ray or gamma ray equipment.

Before selecting radiographic equipment for a task, it must first be determined that radiography will produce the desired test results. Technicians can usually radiograph anything, but the results may not be worth the time, effort and cost. This determination cannot be made until the task has been thoroughly analyzed.

Ideally, there is a best equipment selection for any radiographic test. Practically, most radiography is accomplished by using the equipment available. The equipment lends itself to numerous adaptations and, by knowledgeable choice of film and exposure, any particular equipment can be used for a variety of tasks. For this reason, the capabilities of individual X-ray machines and isotope cameras overlap in many areas of radiographic testing.

Because of its flexibility and ease of operation, X-radiography is often preferred over gamma radiography. Gamma radiography is usually selected for industrial applications that involve the following.

1. High radiation energy requirements.
2. Simultaneous exposure of many test objects.
3. Areas where X-rays cannot be used.
4. Field tests in areas where electrical power is difficult to obtain.

Before the selection of radiographic equipment for a specific test, the radiographer must consider all aspects of the job. Available equipment, the time allocated for the test and the number or frequency of similar object tests are major considerations influencing equipment selection.

ACCESSORY EQUIPMENT

To create a radiograph, only a radiation source, a test object and film are needed. To create a useful radiograph of high quality, additional equipment is required. This equipment, the working tools of the radiographer, includes the following.

1. Diaphragms, collimators and cones.
2. Filters.
3. Screens.
4. Masking material.
5. Image quality indicators.
6. Shim stock.
7. Stepped wedges.
8. Film holders and cassettes.
9. Linear and angular measuring devices.
10. Positioning devices.
11. Identification and orientation markers.
12. Shielding material.
13. Densitometer or stepped wedge comparison film.
14. X-ray exposure charts.
15. Gamma ray exposure charts.
16. Dated decay curves.
17. Film characteristic curves.
18. Table of radiographic equivalence factors.

Diaphragms, Collimators and Cones

Diaphragms, collimators and cones are thicknesses of lead or other dense material, like tungsten, fitted to the tubehead of X-ray equipment or built to contain a gamma ray source. They are designed to limit the area of radiation, as shown in Figure 6.1. They decrease the amount of scatter radiation by limiting the beam to the desired test object area. Many X-ray machines have built in adjustable diaphragms designed so that the beam covers a standard film size area at a fixed distance.

Figure 6.1: Diaphragm, collimator and cone.

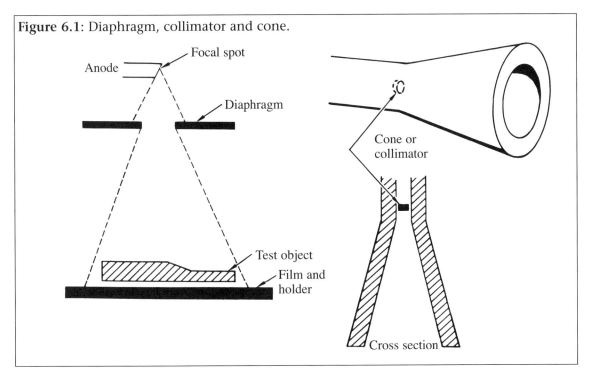

Filters

Filters are sheets of high atomic number metal, usually brass, copper, steel or lead, placed in the X-ray beam at the tubehead, as shown in Figure 6.2. By absorbing the soft radiation of the beam, filters accomplish two purposes: they reduce subject contrast permitting a wide range of test object thicknesses to be recorded with one exposure; and they eliminate scatter caused by soft radiation. Filters are particularly useful in radiography of test objects with adjacent thick and thin sections.

The material and thickness of the test object and its range of thicknesses determine the filter action required. No tables of filter thicknesses are available; however, in radiographing steel, good results have been obtained by using lead filters, 3% of the maximum test object thickness, or copper filters, 20% of the maximum test object thickness.

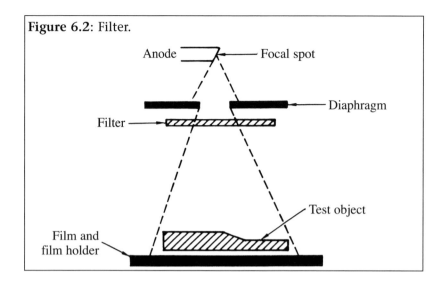

Figure 6.2: Filter.

Anode — Focal spot

Diaphragm

Filter

Test object

Film and film holder

Screens

When an X-ray or gamma ray beam comes in contact with film, less than 1% of the available radiation energy is absorbed by the film in producing an image through photoelectric and compton effect. To convert the unused energy into a form that can be absorbed by film, two types of radiographic screens are used: fluorescent and lead. Lead screens are normally used when high quality is required, whereas fluorescent screens are used when time is a factor. Fluorometallic screens (calcium tungstate with lead) can also be used.

Reduction of backscatter radiation is achieved by placing the film between sheets of lead in the cassette. Normally a 0.005 in. (0.013 cm) thickness is used for the front or top screen, whereas a 0.01 in. (0.025 cm) thickness is used for the back or bottom screen. Screens are used in most radiographic techniques because they reduce the exposure time and improve the quality of the image and increase contrast.

Fluorescent Screens

Fluorescent screens consist of powdered fluorescent material, usually calcium tungstate, bonded to a plastic or cardboard base. When activated by radiation, the fluorescent material emits light in proportion to the amount of radiation available for absorption. The screens are used in pairs with the film placed between them in a film holder.

During exposure, the photographic action on the film is the additive result of the radiation, and the light emitted by the screens impinging on the film. Because the emitted light is diffused, image definition is less sharp when the screens are used. Close contact between the screens and the film must be maintained or the diffused light will cause a blurred, worthless radiograph.

The ratio of an exposure without screens to an exposure with screens, which results in films of similar density, is called the *intensification factor*. Fluorescent screens have a high intensification factor permitting reductions in exposure of the magnitude of 95%. This, however, is the only advantage of using fluorescent screens. Because of their inherent poor image definition characteristic, fluorescent screens are used only in special applications.

Normally, if a radiograph must be taken through concrete looking for rebar or wire position, then fluorescent screens would be of value. Practically, their use is limited to those occasions when a short exposure is required, and the test object conformation permits extensive masking to reduce scattered radiation. Fluorescent screens cause excessive film graininess when exposed to high energy radiation, thus their use is restricted to low energy radiation applications. For high energy applications, they are used with Co-60.

To prevent misleading shadows caused by blocking emitted light during exposure, dirt and dust must be prevented from collecting between the screen and film surfaces. The screens must also be kept free from stains. Their sensitive surfaces must be touched only when necessary and, if cleaning is required, it must be accomplished strictly in accordance with the manufacturer's directions. Also, direct exposure to ultraviolet radiation must be avoided.

Lead Screens

Lead screens are usually constructed of an antimony and lead alloy that is stiffer, harder and more wear resistant than pure lead. The screens are used in pairs on each side of, and in close contact with the film. Depending on the test object and the energy of radiation, the screens may be of varying thicknesses.

The front screen in most applications is thinner than the back screen. Front screens 0.005 in. (0.013 cm) thick and back screens 0.01 in. (0.025 cm) thick are commonly used.

Lead screens are particularly efficient because of their ability to absorb scattered radiation (soft radiation) in addition to increasing the photographic action of the primary radiation on the film. The increased photographic action is a result of the release of electrons from the atoms when acted on by high energy radiation. Energy from the released electrons is readily absorbed by the film emulsion and intensifies film response.

The intensification factor of lead screens is much lower than that of fluorescent screens. During exposure at low energy, it is possible for the front screen absorption effect to be of such magnitude that required exposure is greater than that without screens. However, because of their ability to reduce the effects of scattered radiation and the resulting improvements in contrast and definition of the radiographic image, lead screens are used wherever practical. They are used in almost all gamma ray applications.

To ensure the intensification action of lead screens, they must be kept free from dirt, grease and lint because these materials have high electron absorption qualities and can absorb the intensifying electrons emitted by the screens. The screens may be cleaned with a fine steel wool. The fine abrasion marks caused by gently rubbing with steel wool will have no harmful effects. Deep scratches, gouges, wrinkles or depressions that affect the flatness of the screen surface will cause poor radiographic results.

Masking Material

Masking is the practice of covering or surrounding portions of the test object with highly absorbent material during exposure. Masking reduces the test object exposure in the masked areas, eliminating much scatter. Commonly used masking materials are lead (shown in Figure 6.3), barium clay and metallic shot (shown in Figure 6.4).

When barium clay is used as a mask material, it should be thick enough so that radiation absorption of the clay is appreciably greater than that of the test object; otherwise, the clay will generate noticeable scatter. In any circumstance, the sole purpose of masking is to limit scattered radiation by reducing the area of or about the test object exposed to the primary beam.

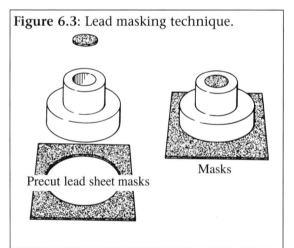

Figure 6.3: Lead masking technique.

Precut lead sheet masks

Masks

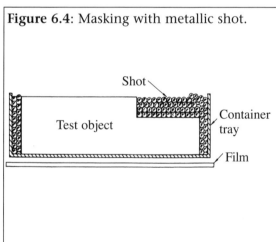

Figure 6.4: Masking with metallic shot.

Shot

Test object

Container tray

Film

Image Quality Indicators

A standard image quality indicator (IQI) is included in every radiograph (some exceptions apply) as a check on the adequacy of the radiographic technique. It is not intended for use in judging the size or in establishing acceptance limits of discontinuities.

Hole type image quality indicators have three holes, one designated the $1T$ hole, the second designated the $2T$ hole, and the third designated the $4T$ hole. Each hole type image quality indicator is identified by an identification number which represents the thickness of the image quality indicator, as shown in Figure 6.5.

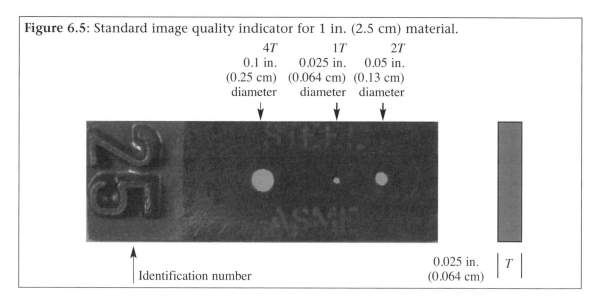

Figure 6.5: Standard image quality indicator for 1 in. (2.5 cm) material.

4T 0.1 in. (0.25 cm) diameter

1T 0.025 in. (0.064 cm) diameter

2T 0.05 in. (0.13 cm) diameter

Identification number

0.025 in. (0.064 cm) | T |

In Figure 6.5, the identification number on an ASME image quality indicator represents a thickness of 0.025 in. (0.064 cm), therefore the 1T hole is 0.025 in. (0.064 cm), the 2T hole is 0.05 in. (0.13 cm) and the 4T hole is 0.1 in. (0.25 cm). In some standards, the selection of image quality indicator will be for a 2% sensitivity which the thickness of the image quality indicator is 2% of the material thickness being radiographed. Standard image quality indicator sizes are listed in Table 6.1.

The image of the outline and desired hole of the image quality indicator on the radiograph is permanent evidence that the radiographic examination achieved the specified sensitivity. The image quality indicator is designed to determine the radiographic quality level, usually referred to as *sensitivity of a radiograph*. Quality levels for image quality indicator sizes are shown in Table 6.2.

Table 6.2: Image quality indicator (IQI) designation, thickness and hole diameters.

Image quality designator designation	IQI thickness, inch (centimeter)	1T hole diameter, inch (centimeter)	2T hole diameter, inch (centimeter)	4T hole diameter, inch (centimeter)
5	0.005 (0.013)	0.01 (0.025)	0.02 (0.051)	0.04 (0.102)
7	0.0075 (0.019)	0.01 (0.025)	0.02 (0.051)	0.04 (0.102)
10	0.01 (0.025)	0.01 (0.025)	0.02 (0.051)	0.04 (0.102)
12	0.0125(0.032)	0.0125(0.032)	0.025 (0.064)	0.05 (0.127)
15	0.015 (0.038)	0.015 (0.038)	0.03 (0.076)	0.06 (0.152)
17	0.0175 (0.044)	0.0175 (0.044)	0.035 (0.089)	0.07 (0.178)
20	0.02 (0.051)	0.02 (0.051)	0.04 (0.102)	0.08 (0.203)
25	0.025 (0.064)	0.025 (0.064)	0.05 (0.127)	0.09 (0.229)
30	0.03 (0.076)	0.03 (0.076)	0.06 (0.152)	0.1 (0.254)
35	0.035 (0.089)	0.035 (0.089)	0.07 (0.178)	0.12 (0.305)
40	0.04 (0.102)	0.04 (0.102)	0.08 (0.203)	0.14 (0.356)
45	0.045 (0.114)	0.045 (0.114)	0.09 (0.229)	0.16 (0.406)
50	0.05 (0.127)	0.05 (0.127)	0.1 (0.254)	0.18 (0.457)
60	0.06 (0.152)	0.06 (0.152)	0.12 (0.305)	0.2 (0.508)
70	0.07 (0.178)	0.07 (0.178)	0.14 (0.356)	0.24 (0.61)
80	0.08 (0.203)	0.08 (0.203)	0.16 (0.406)	0.28 (0.711)
100	0.1 (0.254)	0.1 (0.254)	0.2 (0.508)	0.32 (0.813)
120	0.12 (0.305)	0.12 (0.305)	0.24 (0.61)	0.4 (1.016)
140	0.14 (0.356)	0.14 (0.356)	0.28 (0.711)	0.48 (1.219)
160	0.16 (0.406)	0.16 (0.406)	0.32 (0.813)	0.56 1.422)
200	0.2 (0.508)	0.2 (0.508)	0.4 (1.016)	0.64 (1.626)
240	0.24 (0.61)	0.24 (0.61)	0.48 (1.219)	
280	0.28 (0.711)	0.28 (0.711)	0.56 1.422)	

Table 6.2: Quality levels for hole type image quality indicator sizes.

Equivalent Sensitivity	Quality level	IQI T as percent of T_m	Perceptible hole
0.7%	1-1T	1%	1T
1%	1-2T	1%	2T
1.4%	2-1T	2%	1T
2%	2-2T	2%	2T
2.8%	2-4T	2%	4T
4%	4-2T	4%	2T

Shim Stock

Shim stock is defined as thin pieces of material identical to test object material. Shim shocks are used in radiography of test objects such as welds, where the area of interest is thicker than the nearby test object thickness. Shims are selected so that the thickness of the shim equals the thickness added to the test object (by the weld or a back up bar) in the area of interest, as shown in Figure 6.6.

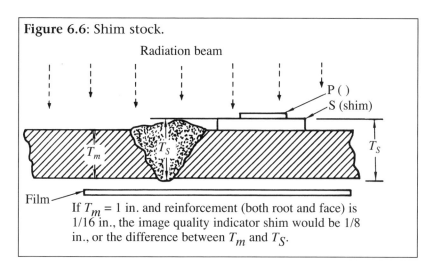

Figure 6.6: Shim stock.

If T_m = 1 in. and reinforcement (both root and face) is 1/16 in., the image quality indicator shim would be 1/8 in., or the difference between T_m and T_S.

The shim is placed underneath the image quality indicator (between the image quality indicator and the test object). In this way, the image of the image quality indicator is projected through a thickness of material equal to the thickness in the area of interest. In use, the length and width of the shim should always be greater than the similar dimensions of the image quality indicator.

Film Holders and Cassettes

Film holders are designed to shield film from light and to protect it from damage. They are made from a variety of materials including rubber and plastic. The holders are flexible and permit molding the film to the contours of the test object, thereby holding the test object to film distance at a minimum.

Cassettes are specially designed, some are two piece hinged, rigid film holders that spring clamp tightly together. These cassettes are of use when flexibility is not required because their clamping action holds screens and film together and firmly in place. Other types of cassettes are more flexible and are usually closed and secure with masking tape or elastic bands.

Linear and Angular Measuring Devices

Correct source to film distance and knowledge of test object thicknesses are required for any radiographic setup. For these measurements, a 6 in. (15 cm) machinist's scale and a tape measure are tools of the radiographer. When a task requires radiography at an angle other than that normal to the plane of the test object, a plumb bob and protractor may be used to determine the correct angular setup.

Positioning Devices

The position of the source (either X-ray or gamma ray), the test object and the film should remain fixed during exposure. With X-ray equipment, the floor, a table or any stable surface may suffice to support the test object. With gamma ray equipment, support of the test object is identical with that of X-ray, and specially designed holders (usually tripods) are used for positioning the source during exposure. Any positioning arrangement complying with safety considerations that does not cause excess scatter radiation is acceptable.

Identification and Location Markers

Location numbers are lead letters and numbers that permit correct correlation of film with the exposure location on the test object. They also prove proper coverage. The test object and the radiograph must be marked so that the test object and its orientation can be identified with the radiograph. This is typically accomplished by affixing lead numbers or letters to, or adjacent to, the test object during exposure and marking the test object in identical fashion with a marking pen or by scribing stencils.

Chalk or paint sticks are commonly used to mark the part or a weld. The lead numbers or letters that are attached with masking tape appear on the radiograph. Comparison of the radiograph with the marked test object eliminates any possibility of wrong identification. In the field, number belts are made by affixing the lead numbers to masking tape. For pipe varying in outside diameter from 2 to 42 in. (5 to 107 cm), the maximum location marker spacing may be determined by the following formula.

Eq. 6.1 $\dfrac{outside\,diameter \times \pi}{number\,of\,films\,used}$

For example, a pipe that has a nominal size of 6 in. (15 cm) has an outside diameter of 6.625 in. (16.8 cm). Three films should be used, based on the *ASME V, Article 2* code specifying a minimum of three films.

Step 1: $6.625\,in. \times \pi = 20.8\,in.$

Step 2: $\dfrac{20.8}{3} = 6.93\,in.$

Step 3: Make a number belt with 6.93 in. (17.6 cm) maximum spacing between the numbers.

Step 4: Place the number belt adjacent to the weld to be radiographed, then mark the placement of the numbers on the test object.

Normally, for pipe, vessels, etc., that are over 42 in. (107 cm) in diameter, a universal number belt is used. Universal number belts have constant spacing between lead numbers with 14 to 15 in. (35.6 to 38 cm) centers. The number belt can be made as large as needed with as many numbers as required. Other options include the use of lead numbers indicating inches from a starting point.

Area Shielding Equipment

The control of scatter radiation is affected only by proper shielding techniques. Areas in which radiography takes place must be adequately protected against both side and backscatter.

In permanent installations, this is accomplished using lead shielded rooms or compartments. When permanent installations are not available, the radiographer uses lead sheets and places them so that areas reached by the primary radiation are shielded. The area immediately beneath or behind the film should always be covered with lead.

Densitometer

The densitometer is an instrument that measures the intensity of light transmitted through film, providing a density value. Two types of densitometers are commercially available: analog and digital meters. Accuracy is a desirable densitometer characteristic, but more important is consistency. A good densitometer, under similar conditions, will give similar readings each time used. Densitometers are very fragile and should not be kept in the darkroom where the chemicals may affect the workings of the instrument.

X-Ray Exposure Charts

X-ray exposure charts, shown in Figure 6.7, illustrate the relationship between material thickness, kilovoltage and exposure. Each chart applies only to a specific set of conditions: a certain X-ray machine; a certain target-to-film distance; a certain manufacture and type of film; certain processing conditions; and the density on which the chart is based.

Exposure charts are useful to determine exposures of test objects of uniform thickness, but should be used only as a guide when radiographing a test object of wide thickness variations. Charts furnished by manufacturers are accurate but only within ±10% because no two X-ray machines are identical. For quality radiography, X-ray exposure charts based on the material most often radiographed, the film most commonly used, and an arbitrarily chosen target-to-film distance are prepared for each X-ray machine in use.

Preparation of an Exposure Chart

To prepare an exposure chart, a series of radiographs is taken using a stepped wedge of the selected test object material. The wedge is radiographed at several different exposures at each of a selected number of kilovoltages. The resultant films are processed in accordance with routine work procedures. Each radiograph will image the wedge as a series of different densities corresponding to the intensity of the X-rays transmitted through the wedge thicknesses.

The radiographer uses a densitometer to locate the desired density on each stepped wedge thickness on the radiograph. At each point the desired density appears, a corresponding value of

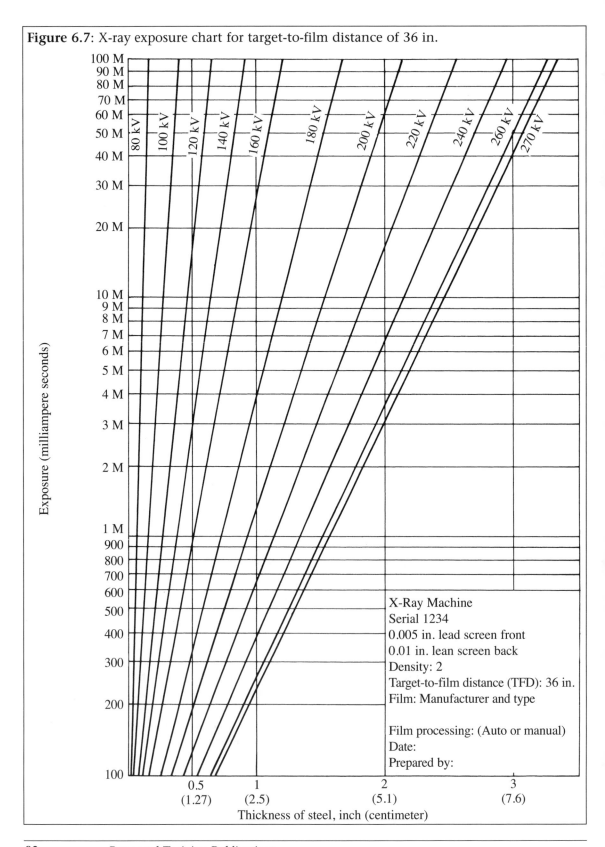

Figure 6.7: X-ray exposure chart for target-to-film distance of 36 in.

Exposure (milliampere seconds)

80 kV 100 kV 120 kV 140 kV 160 kV 180 kV 200 kV 220 kV 240 kV 260 kV 270 kV

X-Ray Machine
Serial 1234
0.005 in. lead screen front
0.01 in. lean screen back
Density: 2
Target-to-film distance (TFD): 36 in.
Film: Manufacturer and type

Film processing: (Auto or manual)
Date:
Prepared by:

Thickness of steel, inch (centimeter)

kilovoltage, exposure and wedge thickness exists. When the desired density does not appear on a radiograph, the correct material thickness for that density is determined by interpolation. The kilovoltage, exposure and material thickness for each of the density points are then plotted on semilog paper. To compress an otherwise overly long scale, the exposure is on the logarithmic scale. Material thickness is designated on the linear scale. The resultant chart will be similar to Figure 6.7 and will be accurate for the particular X-ray machine used.

A second method of preparing an exposure chart requires more calculations, but requires fewer exposures. At each selected kilovoltage, one stepped wedge exposure is made. The densities of each of the wedge thicknesses is measured on each radiograph. Then, an exposure is determined that would have given the desired density under each wedge step by the film characteristic curve. The resultant values of exposure, thickness and kilovoltage are plotted as in the previous method. Use of the film characteristic curve in the preparation of an X-ray exposure chart is shown in the following example.

Example: At 240 kV, a 300 mAs exposure of a steel stepped wedge produced a density of 1.6 under the 1 in. (2.5 cm) thick section of the wedge. At 240 kV, what should the exposure be for a 2.0 density under the 1 in. (2.5 cm) thick section of the wedge, when the film characteristic curve indicates a log relative exposure of 1.8 for a density of 1.6, and 1.91 for a density of 2.0?

The difference between the log relative exposures is 0.11. The antilog of 0.11 is 1.28. Thus, 300 (the exposure for 1.6 density) multiplied by 1.28 will give the exposure for 2.0 density, or $300 \times 1.28 = 384$ mAs.

Film Latitude

Exposure charts can also be prepared to show film latitude, which is defined as the variation in material thickness that can be radiographed with one exposure while maintaining film density within acceptable limits. These limits are fixed by the lowest and highest densities that are acceptable in the finished radiograph. To prepare such an exposure chart, either of the procedures described above are followed, except that both the lowest and the highest acceptable densities are plotted. The result is two curves for each kilovoltage, one representing the lowest and the other the highest acceptable density. For any given exposure and kilovoltage, the range of material thickness capable of being satisfactorily radiographed in a single exposure is shown on the charts as the horizontal difference between the two curves.

Gamma Ray Exposure Chart

A typical gamma ray exposure chart is shown in Figure 6.8. The variables in gamma radiography are the source strength and the source-to-film distance. These are related on the chart to each of three different speed films. By selecting a film, the radiographer can determine the exposure time for a desired image density. Similar to X-ray exposure charts, gamma ray exposure charts are adequate to determine exposures of test objects of uniform thickness but should be used only as a guide when radiographing a test object of wide thickness variation. Charts similar to that shown are available from film manufacturers and are accurate when used with film processed in compliance with the manufacturer's recommendations. The exposure factor shown in the figure is a logarithmic scale of the set of values derived by dividing the product of source strength (gamma ray intensity) and time by the square of the source-to-film distance (inverse square law). The density correction factors were obtained from the film characteristic curves.

Gamma ray exposure charts are easily modified to show latitude. To modify a given chart to reflect highest acceptable density, a curve parallel to the existing curve is drawn. The new curve is displaced vertically above the original by a distance equal to that obtained by applying the density correction factor to the exposure factor at the left edge of the chart. The curve for the lowest acceptable density is drawn in the same manner, but below the original.

An example of this procedure is shown in Figure 6.9, in which the 2.0 density curve for A film is used to develop the 1.5 and 2.5 density curves. The given curve for 2.0 density enters the left edge of the chart at an exposure factor of 6. The correction factor for a density of 1.5 is 0.71. The new curve for 1.5 density enters the left edge at 6×0.71 or 4.26 exposure factor, and continues below and parallel to the 2.0 density curve. Similarly, the curve for 2.5 density enters at 6×1.3 (correction factor for 2.5) or 7.8 and continues above and parallel to the 2.0 density curve. The range of material thickness that can be radiographed in one exposure and result in densities between 1.5 and 2.5 is shown in Figure 6.9 as the horizontal difference between the 1.5 and the 2.5 density curves.

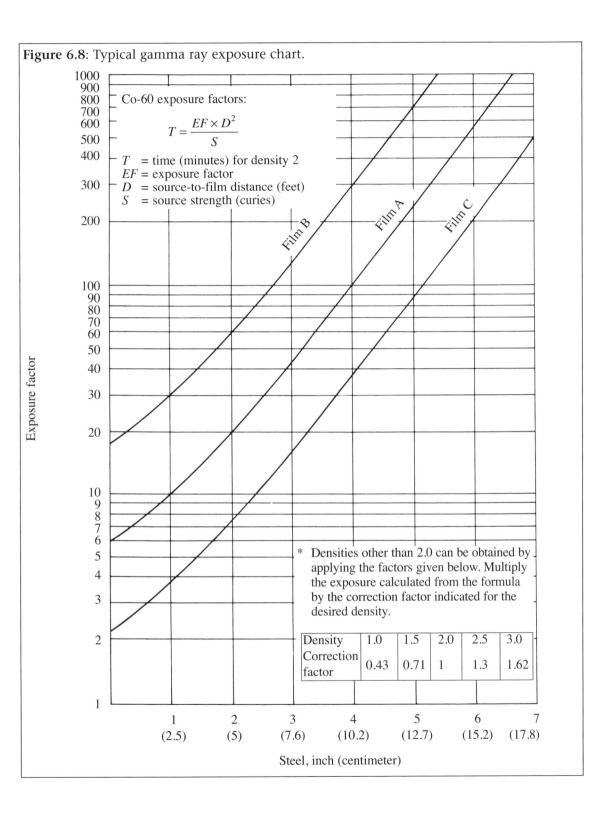

Figure 6.8: Typical gamma ray exposure chart.

Co-60 exposure factors:

$$T = \frac{EF \times D^2}{S}$$

T = time (minutes) for density 2
EF = exposure factor
D = source-to-film distance (feet)
S = source strength (curies)

Film B

Film A

Film C

Exposure factor

* Densities other than 2.0 can be obtained by applying the factors given below. Multiply the exposure calculated from the formula by the correction factor indicated for the desired density.

Density	1.0	1.5	2.0	2.5	3.0
Correction factor	0.43	0.71	1	1.3	1.62

1	2	3	4	5	6	7
(2.5)	(5)	(7.6)	(10.2)	(12.7)	(15.2)	(17.8)

Steel, inch (centimeter)

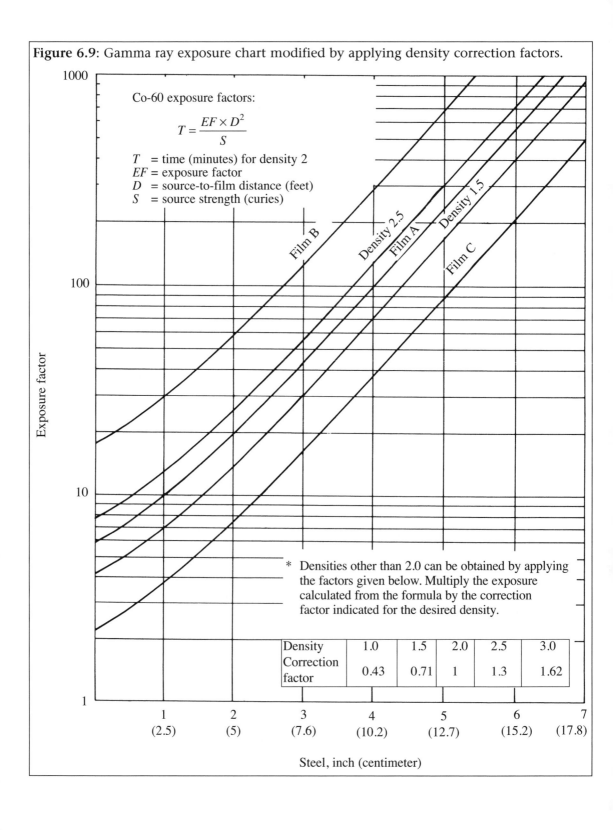

Figure 6.9: Gamma ray exposure chart modified by applying density correction factors.

Co-60 exposure factors:

$$T = \frac{EF \times D^2}{S}$$

T = time (minutes) for density 2
EF = exposure factor
D = source-to-film distance (feet)
S = source strength (curies)

Film B

Density 2.5

Film A

Density 1.5

Film C

* Densities other than 2.0 can be obtained by applying the factors given below. Multiply the exposure calculated from the formula by the correction factor indicated for the desired density.

Density	1.0	1.5	2.0	2.5	3.0
Correction factor	0.43	0.71	1	1.3	1.62

Exposure factor

1 (2.5) 2 (5) 3 (7.6) 4 (10.2) 5 (12.7) 6 (15.2) 7 (17.8)

Steel, inch (centimeter)

Dated Decay Curves

Dated decay curves (Figure 6.10) are usually supplied with radioisotopes. These are computer generated tables of date versus source activity. By use of the curve, the source strength may be determined at any time. Because the source strength must be known before exposure calculations can be made, the decay curve eliminates the necessity of source strength measurement, or calculation, before source use.

When source strength is known, decay curves similar to the one shown are readily prepared by using half life values and plotting the resultant curve on semilogarithmic paper.

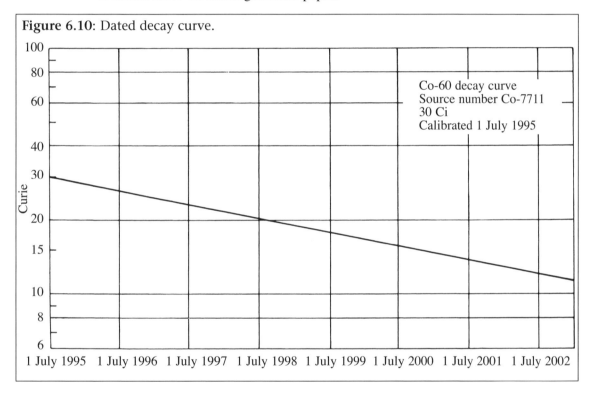

Figure 6.10: Dated decay curve.

Co-60 decay curve
Source number Co-7711
30 Ci
Calibrated 1 July 1995

Film Characteristic Curves

Film curves (discussed in Chapter 4) furnished by manufacturers are accurate and describe the film, but it is always best practice to make one.

Radiographic Equivalent Factors

Most applications of radiation sources are expressed in terms of aluminum or steel thicknesses, as shown in Table 6.3. Radiographic equivalence factors for other commonly used metals are shown in Table 6.4. The values shown are approximate.

In radiographic equivalence tables, aluminum is typically used as the standard metal at 100 kV and below. Steel is the standard at

Table 6.3: Applications of industrial radiation.

X-ray (kV) or isotope	Screens	Approximate practical thickness limits
50	None*	Woods, plastics, thin light metal sections
100	None*	2 in. (5.1 cm) aluminum
150	Lead foil	1 in. (2.5 cm) steel, 4.25 in. (10.8 cm) aluminum
250	Lead foil	2 in. (5.1 cm) steel or equivalent
400	Lead foil	3 in. (7.6 cm) steel or equivalent
1000	Lead foil	5 in. (5.1 cm) steel or equivalent
2000	Lead foil	9 in. (22.9 cm) steel or equivalent
Tm-170	None*	Woods, plastics, light alloys, 0.5 in. (1.27 cm) steel or equivalent
Ir-192	Lead foil	2.5 in (6.4 cm). steel or equivalent
Cs-137	Lead foil	3.5 in. (8.9 cm) steel or equivalent
Co-60	Lead foil	7.5 in. (19.1 cm) steel or equivalent

* Backup screens are recommended in all applications. Lead foil screens as then as 0.001 in. (0.0025 cm) are available with special vacuum pack arrangements that permit screen use with low energy radiation.

Table 6.4: Radiographic equivalent factors.

	X-rays (kV)							Gamma rays		
	50	100	150	220	400	1000	2000	Ir-192	Cs-137	Co-60
Magnesium	0.6	0.6	0.05	0.08				0.22	0.22	0.22
Aluminum	1	1	0.12	0.18				0.34	0.34	0.34
Titanium		8	0.63	0.71	0.71	0.9	0.9	0.9	0.9	0.9
Steel		12	1	1	1	1	1	1	1	1
Copper		18	1.6	1.4	1.4	1.1	1.1	1.1	1.1	1.1
Zinc			1.4	1.3	1.3	1.1	1	1.1	1	1
Brass			1.4	1.3	1.3	1.2	1.2	1.1	1.1	1
Lead			14	12		5	2.5	4	3.2	2.3

higher voltages used with gamma rays. The thickness of the test object is multiplied by the factor shown to obtain an approximate equivalent standard metal thickness.

EXPOSURE VARIABLES

The following sections review and discuss exposure variables as they affect practical radiography techniques.

Movement
Movement of source, test object or film during exposure is always a concern. In high wind areas, care must be taken to ensure that the film or the exposure/guide tube does not move during the

exposure. In X-radiography, permanently installed equipment is designed to remain in the chosen set position, and portable equipment is easily placed so that it does not move. In gamma radiography, the source guide tube probe (source stop) is firmly positioned with clamps, tape, wire, etc. The test object, in either case, is positioned according to its weight, shape and the desired angle of exposure, and the film may be placed and held in position by the tape, magnetics, etc., on the test object when it is not held in position by the weight of the test object. Any means of holding source, test object and film firmly in place is acceptable as long as it does not create scatter radiation problems.

Source Size

Source size is a factor in every radiograph and is a primary consideration in purchasing either X-ray equipment or gamma ray sources. X-ray focal spots vary from 0.5 cm² down to fractions of a millimeter.

The same requirements, and the half life of the radioisotopes under consideration, determine the purchase of a gamma ray source. Generally, the radiographer has available X-ray and gamma ray equipment capable of most radiographic applications, and the problem is how to make an acceptable radiograph with the equipment available. Selection of correct source-to-film distance permits a good radiograph with available equipment, because source (focal spot) size is usually within acceptable dimensions. In gamma radiography, if a smaller source side is required, source manufacturers can produce a smaller dimensioned isotope with a resulting smaller effective source size.

Source-to-Film Distance

Source-to-film distance usually refers to gamma ray equipment, and target-to-film and film-to-focal distance usually refer to X-ray equipment. In the following discussion, the two terms are considered synonymous. In selecting a source-to-film distance, four factors must be considered: source size; test object thickness; test object-to-film distance; and the size of the area of interest.

A longer source-to-film distance will give a closer 1:1 ratio on the film. If the source is too close to the film, the resultant radiograph will have a large penumbral effect (geometric unsharpness).

The penumbral effect is caused by the rays from different points of the source penetrating the test object at different angles (see Figure 2.3). Erroneous densities are caused by the difference in radiation intensity at different points on the test object, which are in turn caused by the difference in distance (inverse square law effect) from the source.

The maximum unsharpness (penumbral effect) that cannot be recognized by the human eye is about 0.02 in. (0.05 cm). Based on this capability of the eye, the following equation can be used to determine a source-to-film distance that gives an acceptable geometric unsharpness.

Eq. 6.2 $D = \dfrac{d \times f}{0.02} + d$

where D is the source-to-film distance, d is the distance from the source side of the test object to the film (test object thickness when the film holder is in contact with the test object) and f is the focal spot size.

A second means of determining source-to-film distance is stated in the commonly used rule: the source-to-film distance should not be less than eight times the test object thickness. Either of these methods of determining source-to-film distance are acceptable for most radiography, but are of little use when thin test objects are radiographed. To radiograph thicker test objects, the radiographer must make sure that the source-to-film distance is sufficient to provide adequate coverage over the area of interest. Another rule of thumb is that the source-to-film distance should be no closer than the maximum diagonal dimension of the film (detector) holder (cassette).

Usually, a source-to-film distance is selected long enough for all anticipated test object exposures, and exposure charts are constructed on the basis of that distance.

Film Contrast, Speed and Graininess

Film characteristics are discussed in Chapter 4. With most industrial used films, the same degree of contrast is obtainable regardless of the speed of the film because the characteristic curves of the different speed films are similar in shape. The degree of resolution (sharpness) required in the radiograph determines the speed of film that is acceptable. The time saved, economics and consideration of fast film is secondary to the desired resolution. Fast film is rarely used for high quality work. Fast film is always used with concrete and very thick parts.

Controlling Scatter Radiation

Lead screens are widely used because of their intensifying action to reduce exposures and their scatter absorption capability. They are available in a wide range of thicknesses, and extremely thin lead screens in vacuum pack film holders are successfully used in radiography of thin test objects. Exposure charts should be based on lead screen exposures. Because of the loss of sharpness that accompanies their use, calcium tungstate screens should be used only to reduce exposure times.

Scatter radiation can never be eliminated, but its effects can be lessened by limiting the amount of scatter and by further limiting the scatter reaching the film. Filters placed between the source and the test object absorb many of the scatter producing soft rays of the beam and may be effective with X-ray equipment. They are not required in gamma radiography because of the high energy of gamma ray emissions. Lead screens absorb both internal scatter (front screen) and backscatter (back screen); collimators, cones and diaphragms reduce side and backscatter by limiting the beam to the area of interest.

Scatter, generated by the test object (internal scatter), is reduced by limiting the area of the test object exposed to the beam. Masks of lead, barium clay, metallic shot or other absorbent materials are used to shield portions of the test object or areas surrounding the test object. The principle of masking is the same as that of filtering, except that filters are designed to absorb only soft rays, whereas masks absorb the soft rays and many of the higher energy rays.

Shields serve to limit scatter radiation by absorbing rays that might otherwise strike walls, floors or objects that would generate scatter. Shields usually consist of lead sheets, in some convenient handling form, placed in positions of most scatter reduction benefit. It is particularly important that areas immediately below or behind the film be shielded to absorb backscatter. Permanent radiography installations usually include lead lined rooms; but, in all other applications, control of scatter by area shielding is required.

Kilovoltage, Milliamperage and Time

Kilovoltage, milliamperage and time are exposure factors in X-radiography. X-ray exposure charts, shown in Figure 6.7, describe the correct value of each factor for certain applications. A combination of the information contained in an exposure chart and the information contained in a table of radiographic equivalents (Table 6.5) results in determination of proper exposure values for X-ray of material other than that shown on the chart.

Application of the inverse square law to exposure chart information results in correct exposure values for different source-to-film distances. Combining exposure chart information with information obtained from film characteristic curves results in correct exposure values for various speed films. Because milliamperage and time are reciprocal functions and milliamperage is limited by equipment capability, required exposure time is usually determined by the equipment used.

Source Energy, Strength and Time

Source energy, source strength and time are exposure factors in gamma radiography. Gamma ray exposure charts, shown in Figure 6.8, describe the proper value of each under certain conditions. Source energy (wavelength of the emitted waves) is a function of the radioisotope source and remains constant. Source strength is a time decay function of the radioisotope and must be known at the time of exposure. Because source strength and time are reciprocal functions, the length of exposure time is determined by the source strength.

Gamma ray exposure chart information combined with a table of radiographic equivalents (Table 6.5) results in determination of correct exposure values for material not shown on the chart. The exposure formula accompanying most gamma ray exposure charts allows for application of the inverse square law, and the remainder of the information on the chart permits selection of a source-to-film distance most suited for the immediate task.

Absorption and Contrast

Test object absorption and resulting subject contrast on the radiograph are variables of the radiographic process that are difficult to control. They determine the radiographer's control or setting of the other variables encountered in the making of a radiograph. 7Knowledge of the composition of the material is a benefit when calculating the exposure time. A 1 in. (2.5 cm) plate of nickel chromium alloy will not absorb the same amount of radiation as would a 1 in. (2.5 cm) plate of low thermal expansion superalloy, such as aluminum.

EXPOSURE CALCULATIONS

The following examples of exposure calculations illustrate the equipment and film information available to the radiographer. The referenced figures are located together for ease in following the examples. The equipment used in the examples consists of a portable X-ray machine whose characteristics are shown in Figure 6.11, a permanently installed X-ray machine whose characteristics are shown in Figure 6.12, and an isotope camera containing an Ir-192 source whose decay curve is shown in Figure 6.13. The film used in the examples is of six different types: types I, II and III typical of one manufacturer and types A, B and C typical of another.

Characteristic curves for types I, II and III film are shown in Figure 6.14. An Ir-192 exposure chart for types A, B and C film is shown in Figure 6.15. Figure 6.16 details the maximum permissible voltage for specified material thicknesses that will obtain 2% sensitivity in the finished radiograph. It also details the permissible thickness range of material for gamma radiography that will obtain 2% sensitivity.

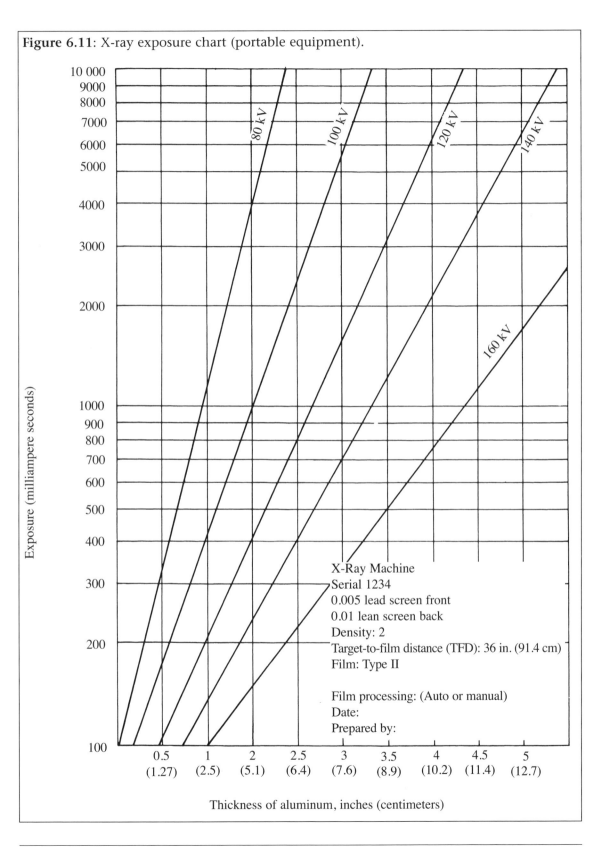

Figure 6.11: X-ray exposure chart (portable equipment).

X-Ray Machine
Serial 1234
0.005 lead screen front
0.01 lean screen back
Density: 2
Target-to-film distance (TFD): 36 in. (91.4 cm)
Film: Type II

Film processing: (Auto or manual)
Date:
Prepared by:

Exposure (milliampere seconds)

Thickness of aluminum, inches (centimeters)

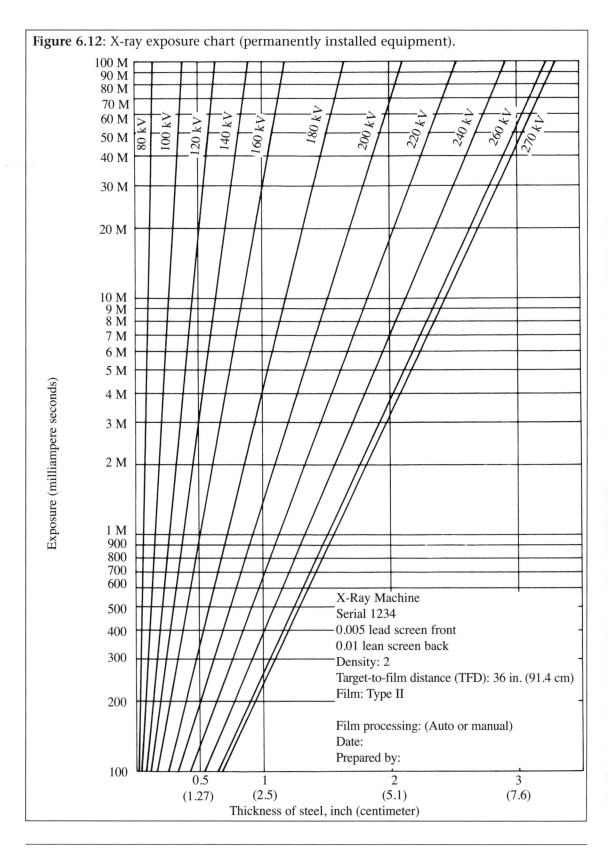

Figure 6.12: X-ray exposure chart (permanently installed equipment).

X-Ray Machine
Serial 1234
0.005 lead screen front
0.01 lean screen back
Density: 2
Target-to-film distance (TFD): 36 in. (91.4 cm)
Film: Type II

Film processing: (Auto or manual)
Date:
Prepared by:

Exposure (milliampere seconds)

Thickness of steel, inch (centimeter)

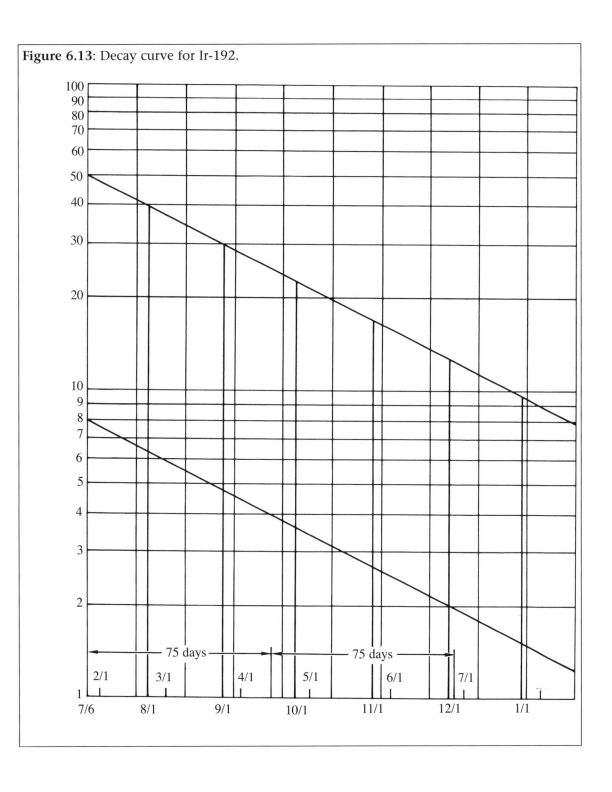

Figure 6.13: Decay curve for Ir-192.

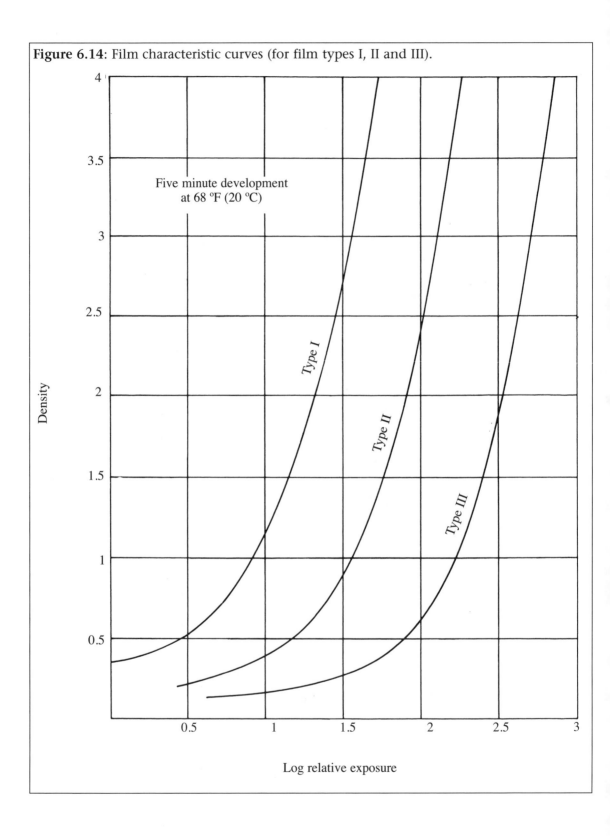

Figure 6.14: Film characteristic curves (for film types I, II and III).

Five minute development
at 68 ºF (20 ºC)

Type I

Type II

Type III

Density

Log relative exposure

Figure 6.15: Exposure chart for Ir-192 with film types A, B and C.

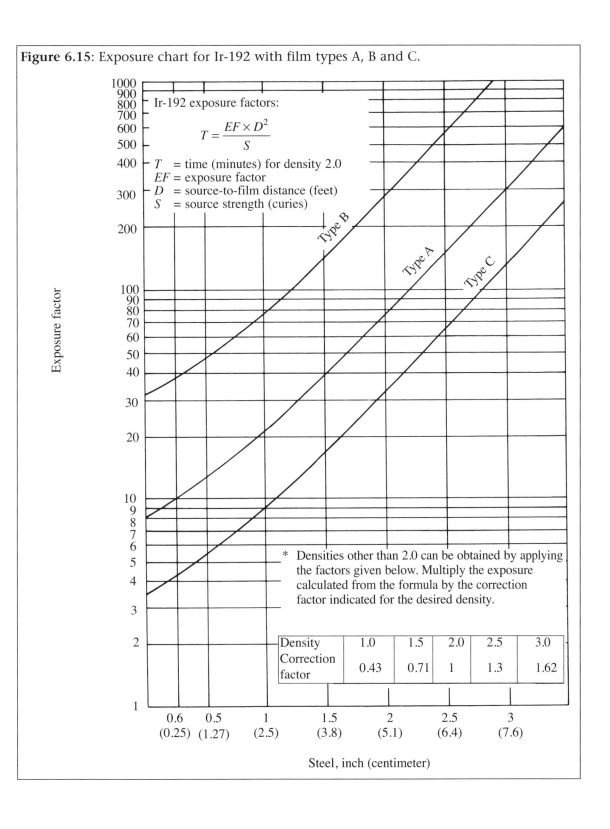

Ir-192 exposure factors:

$$T = \frac{EF \times D^2}{S}$$

T = time (minutes) for density 2.0
EF = exposure factor
D = source-to-film distance (feet)
S = source strength (curies)

Exposure factor

Type B

Type A

Type C

* Densities other than 2.0 can be obtained by applying the factors given below. Multiply the exposure calculated from the formula by the correction factor indicated for the desired density.

Density	1.0	1.5	2.0	2.5	3.0
Correction factor	0.43	0.71	1	1.3	1.62

| 0.6 (0.25) | 0.5 (1.27) | 1 (2.5) | 1.5 (3.8) | 2 (5.1) | 2.5 (6.4) | 3 (7.6) |

Steel, inch (centimeter)

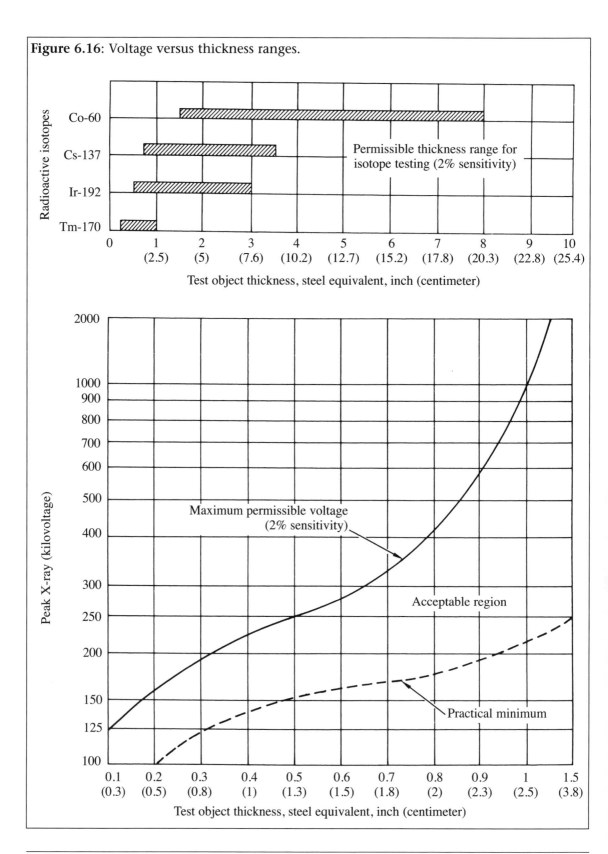

Figure 6.16: Voltage versus thickness ranges.

Radioactive isotopes

Co-60

Cs-137

Ir-192

Tm-170

Permissible thickness range for
isotope testing (2% sensitivity)

0   1   2   3   4   5   6   7   8   9   10
(2.5) (5) (7.6) (10.2) (12.7) (15.2) (17.8) (20.3) (22.8) (25.4)

Test object thickness, steel equivalent, inch (centimeter)

Peak X-ray (kilovoltage)

2000

1000
900
800
700
600

500

400

300

250

200

150

125

100

Maximum permissible voltage
(2% sensitivity)

Acceptable region

Practical minimum

0.1   0.2   0.3   0.4   0.5   0.6   0.7   0.8   0.9   1   1.5
(0.3) (0.5) (0.8) (1) (1.3) (1.5) (1.8) (2) (2.3) (2.5) (3.8)

Test object thickness, steel equivalent, inch (centimeter)

Example 1: A 2 in. (5 cm) thick test object of aluminum is to be radiographed. Using the portable X-ray equipment (Figure 6.11), determine the exposure for a sensitivity of 2% and a density of 3.3 with type II film.

> **Step 1**: As seen in Figure 6.11, 2 in. (5 cm) of aluminum can be radiographed with type II film at a target-to-film distance of 36 in. (91 cm) for a density of 2.0 as follows.
>
> 1. At 100 kV: 1000 mAs.
> 2. At 120 kV: 400 mAs.
> 3. At 140kV: 230 mAs.
> 4. At 160 kV: 160 mAs.
>
> **Step 2**: As shown in Figure 6.14, the log relative exposure with type II film for a 2.0 density is 1.91, and for 3.3 density 2.18. The difference between the log relative exposures is 0.27. The antilogarithm of 0.27 is 1.83. Therefore, to obtain the exposure for 3.3 density, the exposure for 2.0 density is multiplied by 1.83.
>
> **Step 3**: Following Step 1, the exposure for 2.0 density at 140 kV is 230 mAs. Thus, at 140 kV, an exposure of 421 mAs (230 × 1.83) will result in a radiograph of 2% sensitivity and 3.3 density.

Example 2: In Example 1, determine the exposure required with type III film.

> **Step 1**: As shown in Figure 6.14, the log relative exposure for 3.3 density with type II film is 2.18, and with type III film is 2.76. The difference between the log relative exposures is 0.68. The antilog of 0.68 is 4.78. Therefore to obtain the exposure with type III film, the exposure for type II film is multiplied by 4.78.
>
> **Step 2**: From Example 1, the exposure for type II film is 421 mAs. Thus, an exposure of 2012 mAs (732 × 4.78) will result in a radiograph of 2% sensitivity and 3.3 density with type III film.

Example 3: In Examples 1 and 2, the exposure was based on a target-to-film distance of 36 in. (91 cm). The rule previously given states that the target-to-film distance should not be less than eight to ten times the thickness of the test object. Based on this rule, a 20 in. (51 cm) target-to-film distance is selected because of the possible saving in time. Determine the exposure for 2% sensitivity and 3.3 density at this target-to-film distance for types II and III film.

Step 1: The inverse square law states that the intensity varies inversely with the square of the distance. Thus, the exposure at a 20 in. (51 cm) target-to-film distance for 2% sensitivity and 3.3 density is 400/1296 ($20^2/36^2$) of the exposure at a 36 in. (91 cm) target-to-film distance for 2% sensitivity and 3.3 density.

Step 2: From Example 1, the exposure for type II film at a 36 in. (91 cm) target-to-film distance is 421 mAs.

Step 3: Thus, an exposure of 130 mAs (421 × 400/1296) will result in a radiograph of 2% sensitivity and 3.3 density with type II film at a target-to-film distance of 20 in. (51 cm).

Step 4: From Example 2, the exposure for type Ill film at a 36 in. (91 cm) target-to-film distance is 3500 mAs.

Thus, an exposure of 621 mAs (2012 × 400/1296) will result in a radiograph of 2% sensitivity and 3.3 density with type III film at a target-to-film distance of 20 in. (51 cm). Note: The procedures of Examples 1 through 3 may be followed to X-ray the test object using the equipment described in Figure 6.12, and the radiographic equivalence factors listed in Table 6.4.

Example 4: The test object of Example 1 must be radiographed with Ir-192. Fifty-five days have passed because the source shown in Figure 6.13 was established at 50 Ci. Using this source, determine the exposure with type A film to obtain 2% sensitivity and 3.3 density.

Step 1: As shown in Figure 6.13, the 50 Ci source after 55 days decayed to 30 Ci.

Step 2: As in Example 3, select a source-to-film distance of 20 in. (51 cm).

Step 3: As shown in Table 6.4, the radiographic equivalence factor for aluminum, when using Ir-192, is 0.34. Thus, with Ir-192, 2 in. (5 cm) of aluminum are equivalent to 0.68 in. of steel (2 × 0.34).

Step 4: As shown in Figure 6.16, the lower level of the permissible thickness range for Ir-192 radiography with 2% sensitivity is 0.5 in. (1.3 cm) of steel. Therefore, 2 in. (5 cm) of aluminum or 0.68 in. (1.7 cm) of steel can be radiographed with Ir-192 with 2% sensitivity.

Step 5: Using the equation from Figure 6.15, calculate *T* in minutes.

$$T = \frac{EF \times D^2}{S} \qquad \left(\frac{15 \times 1.7^2}{30} = 1.45\,\text{min}\;or\;87\,\text{s} \right)$$

Thus, an exposure time of 84 s will result in a radiograph of 2% sensitivity and 2.0 density.

Step 6: Figure 6.15 does not give the correction factor for 3.3 density. To obtain a density of 3.3, the exposure for 2.0 density must be increased by the ratio of 2.8:1.5 (1.87).

Step 7: From Step 5, the exposure time for 2 density is 87 s. Thus, an exposure time of 163 s will result in a radiograph of 2% sensitivity and 3.3 density.

Note: In Examples 1 and 2, the film characteristic curves were plotted on a log relative scale, and it was necessary to determine the antilog of the log relative exposure difference between any two exposures to calculate required exposure change. In this example, the film characteristic curve is plotted on a logarithmic scale in actual exposure values, and calculation of required exposure changes is a matter of applying the ratio between any two exposures.

Example 5: The steel test object shown in Figure 6.17a is to be radiographed. Required sensitivity is 2%, maximum acceptable density is 3.3, minimum is 2.0. Using the X-ray equipment described in Figure 6.12, determine if a radiograph of acceptable sensitivity and densities can be made with a single exposure of type II film.

Figure 6.17: Steel test objects.

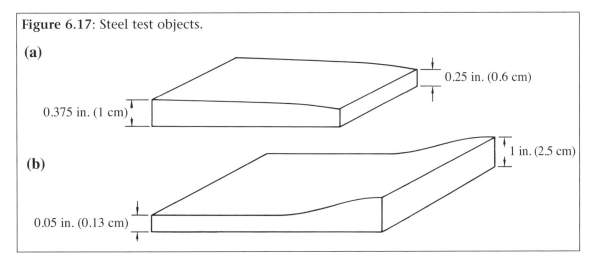

(a)

0.375 in. (1 cm)

0.25 in. (0.6 cm)

(b)

1 in. (2.5 cm)

0.05 in. (0.13 cm)

Step 1: From Figure 6.16, the maximum permissible voltage for 2% sensitivity with 0.25 in. (0.6 cm) of steel is 170 kV, and the practical minimum voltage for 0.375 in. (1 cm) of steel is 135 kV. Therefore, only 140 and 160 kV curves of Figure 6.12 are considered.

Step 2: As shown in Figure 6.12, with type II film at a target-to-film distance of 36 in. (91 cm), 0.25 in. (0.6 cm) and 0.375 in. (1 cm) of steel may be radiographed for a density of 2.0 as shown in Table 6.5.

Table 6.5: Exposure chart for example 5, step 2 and example 7, step 2.		
Exposure steel thickness	**140 kV**	**160 kV**
0.25 in. (0.6 cm)	330 mAs	170 mAs
0.375 in. (1 cm)	1000 mAs	400 mAs

Step 3: As shown in Figure 6.14, the log relative exposure with type II film for a 2.0 density is 1.91, and for 3.3 density is 2.18. The difference between the log relative exposures is 0.27. The antilog of 0.27 is 1.83. Therefore, to obtain the exposure for 3.3 density, the exposure for 2.0 density is multiplied by 1.83.

Step 4: From Step 2, the exposure of 0.25 in. (0.6 cm) of steel for 2 density at 140 kV is 330 mAs, and at 160 kV is 170 mAs. Thus, an exposure of 604 mAs (330 × 1.83) at 140 kV, and 311 mAs (170 × 1.83) at 160 kV will result in radiographs of 3.3 density.

Step 5: Therefore, exposures within the acceptable density range are shown in Table 6.6.

Table 6.6: Exposure chart for example 5, step 5.		
Exposure steel thickness and density	**140 kV**	**160 kV**
0.25 in. (0.6 cm), 3.3 density	604 mAs	311 mAs
0.375 in. (1 cm), 2.0 density	1000 mAs	400 mAs

With 140 kV, any exposure more than 604 mAs will result in a density greater than 3.3 at the thin portion of the test object, and any exposure less than 1000 mAs will result in a density of less than 2.0 at the thick portion of the test object. The same relative conditions hold true with 160 kV. It is impossible to obtain a radiograph of acceptable sensitivity and densities with a single exposure of type II film.

Example 6: The steel test object shown in Figure 6.17b must be radiographed with Ir-192. The available source measures 30 Ci. Required sensitivity is 2%, maximum acceptable density is 3.3 and minimum is 2.0. Determine if a radiograph of acceptable sensitivity and densities can be made with a single exposure of type A film.

Step 1: Refer to Figure 6.16. The lower level of the permissible thickness range for Ir-192 radiography with 2% sensitivity is 0.5 in. (1.3 cm) of steel. Therefore, the test object can be radiographed with Ir-192 with 2% sensitivity.

Step 2: Refer to Figure 6.15. The exposure factor for 0.5 in. (1.3 cm) steel for 2.0 density with type A film is 12.5, and 1 in. (2.5 cm) of steel is 21.

Step 3: Figure 6.15 does not give the correction factor for 3.3 density. The exposure with type A film for 2.0 density is 1.5 R, and for 3.3 density 2.8 R. Therefore, to obtain a density of 3.3, the exposure for 2.0 density must be increased by the ratio of 2.8:1.5 (1.87).

Step 4: From Step 2, the exposure factor of 0.5 in. (1.3 cm) of steel for 2.0 density with type A film is 12.5. Thus an exposure factor of 23.4 (12.5 × 1.87) will result in a radiograph of 3.3 density.

Step 5: Because an exposure factor of 21 will result in 2.0 density through the thicker 1 in. (2.5 cm) portion of the test object, and an exposure factor of 23.4 will result in 3.3 density through the thinner portion of the test object, any exposure factor between 21 and 23.4 will result in a radiograph of acceptable sensitivity and density with a single exposure of type A film.

Double Film Exposures

The test object of Example 5, which could not be radiographed satisfactorily with a single exposure on one film, may be radiographed by using two exposures, one for the thicker portion of the test object and one for the thinner portion. However, the test object also may be radiographed with a single exposure using two films of different speeds with consequent savings in time. In this double film technique, the two films are placed in the same holder and exposed simultaneously.

This is practical because the absorption of radiation by film is so slight that the effect of the radiation on either of the two films is, for practical purposes, similar to that of a single film exposure.

The exposure ratio between the films used in the double film technique determines the range of test object thickness that can be radiographed with acceptable density. The ratio of exposure between fine (medium speed) and extra fine (slow speed) film ranges from

1:3 to more than 1:4, dependent on the particular film characteristics as set by the manufacturer. Because of this high ratio, calculations for the double film technique are based on an exposure for maximum acceptable density through the thicker portions of the test object, recorded on the faster of the two films; and an acceptable density through the thinner portions of the test object, recorded on the slower film. (Fast, coarse grain film is seldom used for this purpose.)

Example 7: The steel test object shown in Figure 6.17a is to be radiographed. Required sensitivity is 2%, maximum acceptable density is 3.3, minimum is 2.0. Using the X-ray equipment described in Figure 6.12, determine if radiographs of acceptable density can be made with types II and III film and the double film technique.

Step 1: Refer to Figure 6.16. The maximum permissible voltage for 2% sensitivity with 0.25 in. (0.6 cm) of steel is 170 kV, and the practical minimum voltage for 0.375 in. (1 cm) of steel is 135 kV. Therefore, only the 140 and 160 kV curves of Figure 6.12 are considered.

Step 2: Refer to Figure 6.12. With type II film at a target-to-film distance of 36 in. (91 cm), 0.25 in. (0.6 cm) and 0.375 in. (1 cm) of steel may be radiographed for a density of 2, as shown in Table 6.6.

Step 3: Refer to Figure 6.14. The log relative exposure with type II film for 2.0 density is 1.91, and for 3.3 density is 2.18. The difference between the log relative exposures is 0.27. The antilog of 0.27 is 1.83. Therefore, to obtain the exposure for 3.3 density, the exposure for 2.0 density is multiplied by 1.83. The log relative exposure with type III film for 2.0 density is 2.53. The difference between the log relative exposures for 2.0 density with types II and III film is 0.62. The antilog of 0.62 is 4.17. Therefore, to obtain the exposure for 2.0 density with type III film, the exposure for type II film is multiplied by 4.17.

Step 4: The exposure with type II film of 0.25 in. (0.6 cm) of steel for 2.0 density at 140 kV is 330 mAs and at 160 kV is 170 mAs. Thus, an exposure of 1376 mAs (330×4.17) at 140 kV, and 709 mAs (170×4.17) at 160 kV will result in radiographs of 2.0 density with type III film.

Step 5: The exposure of 0.375 in. (1 cm) steel for 2.0 density at 140 kV is 1000 mAs, and at 160 kV is 400 mAs. Thus, an exposure of 1830 mAs (1000×1.83) at 140 kV and 732 mAs (400×1.83) at 160 kV will result in radiographs of 3.3 density with type II film.

Step 6: Therefore exposures within the acceptable density range are shown in Table 6.7.

Table 6.7: Exposure chart for example 7, step 6.

Exposure steel thickness	140 kV	160 kV
0.25 in. (0.6 cm), 2.0 density, type III film	1376 mAs	709 mAs
0.375 in. (1 cm), 3.3 density, type II film	1830 mAs	732 mAs

Because with 140 kV any exposure more than 1376 mAs and less than 1830 mAs will result in radiographs with a density greater than 2.0 at the thin portion of the test object on type III film and a density less than 3.3 at the thick portion of the film on type II film, radiographs of acceptable density can be made with one exposure at 140 kV. The 709 mAs exposure for the thin portion of the test object and the 732 mAs exposure for the thick portion of the test object at 160 kV are about equal. Exposing the test object to 160 kV at an exposure between 709 and 732 mAs will result in radiographs of acceptable density.

Note: the calculations in the examples above and in most radiography are close approximations and not exact values. Minor variances in film, in equipment performance and in measurement capabilities (time, distance, density) do not permit exact calculations.

Radiographic Slide Rules

The principles of exposure calculation illustrated in the previous examples may be applied to almost any exposure. Slide rules, calculators and similar devices designed to assist the radiographer in calculating exposures are simply handy arrangements of the information contained in film characteristic curves, exposure charts, radiographic equivalence tables and application of the inverse square law. These devices are reasonably accurate, and the information obtained from them may be relied on, providing the user understands the principles illustrated in the examples.

RADIOGRAPHIC APPLICATIONS

The exposure arrangements discussed in the following paragraphs and illustrated in Figures 6.18 through 6.40 are commonly used, and application of these principles permits radiography of most test objects. Except where otherwise specified, any of these arrangements may be used with either X-ray or gamma ray equipment. The basic principles of film density and contrast as related to source-to-film distance, source energy and exposure apply to each of the arrangements.

Radiography of Welds

Tube Angulation

Before setup and exposure of any weld configuration, the radiographer must know the joint preparation, weld penetration standard and fusion lines in order to set the tube angulations (direction of the beam) and the resultant incident beam propagation path.

Incident Beam Alignment

The incident beam is the central beam of the radiation field. It is the effective focal spot size, projected in straight lines, to the center of the area of interest.

Discontinuity Location

Sometimes it is essential to locate discontinuities in exceptionally thick test objects where the depth of the discontinuity must be known to remove a minimum amount of material from the nearest side. Correctly locating and removing the discontinuity will save manufacturing and radiography time and conserve both manufacturing and radiographic materials. Depth cannot, as a rule, be judged by radiography. There are several methods such as stereoradiography and the parallax method that can be used to judge the depth, but these are not available in a typical field setup.

Critical and Noncritical Criteria

The radiographer must know the acceptance criteria and area of interest of every test object before any function is performed in the radiographic process. The radiographer must decide which film will give the least and/or highest sensitivity, the radiographer must determine the distance and angle to give the least amount of distortion and the radiographer must determine the number of exposures necessary to provide complete coverage of the area of interest. Typical radiographic requirements are as follows.

1. Extent and distribution of radiographic testing for initial and subsequent welds.
2. Specific welds to be examined.
3. Numerical sequence of welds to be examined.
4. Radiographic standards to apply to each weld.

Improper Interpretation of Discontinuities

To properly interpret discontinuities, all factors of the manufacturing or welding process must be correctly applied, noted and known by personnel involved in overall and final evaluation. Figures 6.18 to 6.40 illustrate some typical weld configurations and correct and incorrect positions.

Elimination of Distortion

Observing the proper geometry of exposure will minimize distortion by showing the image in the proper perspective. The source should typically be perpendicular to the surface of an object and plane of the film (detector).

Proper Identification and Image Quality Indicator Placement

Image quality indicators are added to a test object to show sensitivity. They can also serve other purposes, such as image orientation with elliptical views.

The identification plate is used to identify each and every individual exposure. Information such as test object number, X-ray control number, weld number, area number, date of exposure and any other pertinent information is provided. The identification plate can also serve the purpose of orientation. For large areas that require more than one view for coverage, lead location markers are used to correlate the radiograph to the location on the weld or component.

Radiography of Welded Flat Plates

Figure 6.18 illustrates flat weld areas. This type of weld is easily radiographed because its area of interest is clearly defined in its length, width and thickness. Subject contrast is small and exposure calculations are relatively simple.

To ensure the correct degree of sensitivity, the radiographer must select the proper image quality indicator and sufficient shim stock (if hole type image quality indicators are used) so that the image of the image quality indicator is a true representation of sensitivity for the thickness of the test object at the weld area.

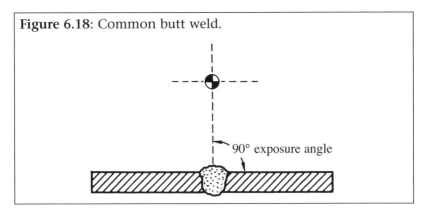

Figure 6.18: Common butt weld.

90° exposure angle

Radiography of Welded Corner Joints

Figures 6.19 to 6.21 are illustrative of correct and incorrect X-ray source placement of a corner joint. Proper criteria should cover all weld configurations to show them to the best advantage on the film. The deciding factors are welding standards, joint configuration and design stress.

Figure 6.19 shows correct tube angle, test object placement and joint alignment. Figure 6.20 shows correct tube angle, but tube and part have been placed in such a position that the unfused area will appear on the film resulting in incorrect interpretation.

Figure 6.21 illustrates 100% joint penetration. The X-ray tube angle of 45° is correct. The film and joint must be placed perpendicular to the tube aperture.

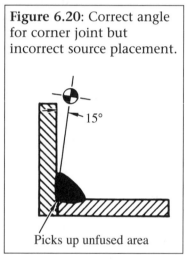

Figure 6.19: Correct angle for corner joint and correct source placement.

Figure 6.20: Correct angle for corner joint but incorrect source placement.

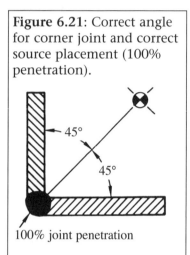

Figure 6.21: Correct angle for corner joint and correct source placement (100% penetration).

Single Wall Radiography of Tubing

Figure 6.22 shows an example of a single wall radiograph that should be used when possible. This is true of flat objects as well as circular objects. Factors relating to all single wall applications are as follows.

All circular test objects should be numbered in a clockwise direction. Lead numbers should be placed adjacent to the weld and at least 0.125 in. (0.3 cm) from the heat affected zone.

A good method to retain identification is to electrolytically etch the numbers or to use metal impression stamps if the specifications permit. Lead numbers, when used, should be taped or otherwise temporarily affixed to the test object. Numbers should be placed on both sides of the identification plate and image quality indicator.

In laying out a circumferential weld for the least amount of geometrical distortion, calculate (on both sides of each area) the points at which the greatest visual circumferential changes take place. Deduct about 10% from both sides to allow for distortion.

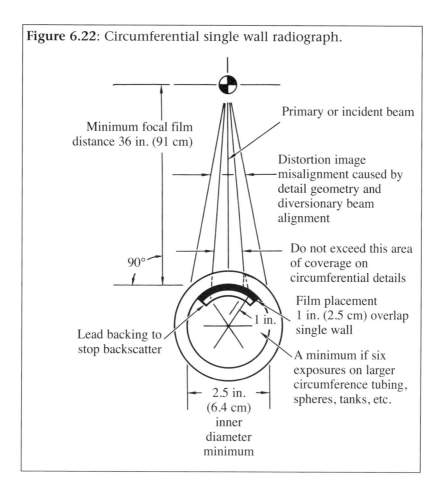

Figure 6.22: Circumferential single wall radiograph.

Minimum focal film distance 36 in. (91 cm)

Primary or incident beam

Distortion image misalignment caused by detail geometry and diversionary beam alignment

Do not exceed this area of coverage on circumferential details

90°

Film placement 1 in. (2.5 cm) overlap single wall

1 in.

Lead backing to stop backscatter

A minimum if six exposures on larger circumference tubing, spheres, tanks, etc.

2.5 in. (6.4 cm) inner diameter minimum

Another good method for discontinuity location and area orientation is to place lead arrows with adhesive backs in the center and at the ends of each area. These arrows must remain on the test object until the film has been interpreted, and then removed. All lead tables and secondary radiation backing should be covered to protect the test object where contamination is a concern.

Double Wall Radiography of Tubing

Figures 6.22 and 6.24 illustrate geometric principles, minimum distortion and orientation related to double wall radiography applications.

Tubing up to 3.5 in. (9 cm) Outside Diameter

As seen in Figure 6.23, the source side image quality indicator is selected based on the total material the beam of radiation passes through, unless otherwise specified. Lead area numbers and the source side image quality indicator should be on the side away from the angled beam so they will not superimpose in the weld area.

Two 90° opposing shots should be taken to provide full coverage on elliptical views. The angle shots are required for discontinuity orientation.

Figure 6.23: Double wall radiography with tube inside diameter equal to or greater than 1.2 in (3 cm).

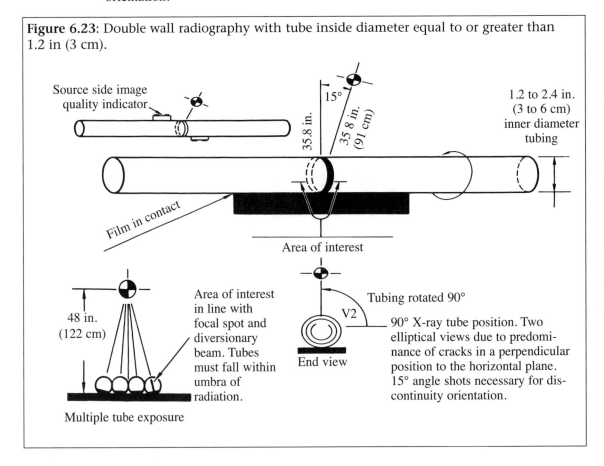

Film should be in contact to minimize distortion and unsharpness of the image.

The lower left of Figure 6.23 illustrates a method for exposing more than one tube assembly on a single exposure. Note that the areas of interest are adjacent to the film and angled to align with the focal spot. To minimize distortion, it is best to use a focal film distance of 48 in. (122 cm) or more.

Radiography of Closed Spheres

The radiographic applications for a closed sphere are shown in Figure 6.25. The applications are similar to those for double wall tubing. The image quality indicators must be placed on a block of similar material to show total thickness of the double wall. In this case the area numbers may be face up with the identification plate if desired, because the area can be more easily oriented.

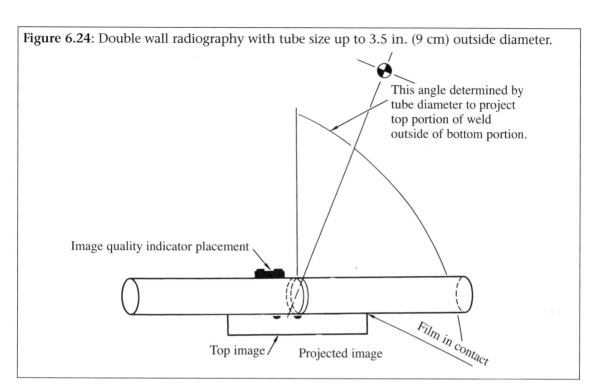

Figure 6.24: Double wall radiography with tube size up to 3.5 in. (9 cm) outside diameter.

This angle determined by tube diameter to project top portion of weld outside of bottom portion.

Image quality indicator placement

Top image / Projected image

Film in contact

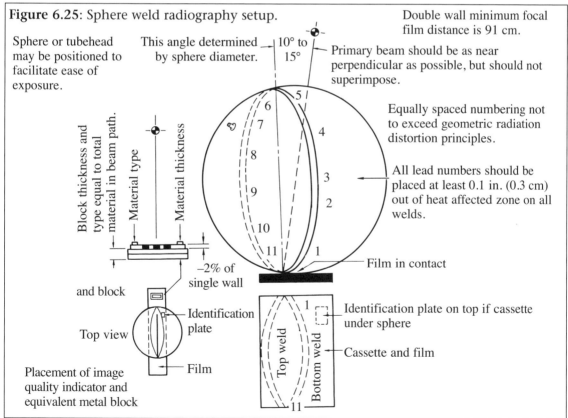

Figure 6.25: Sphere weld radiography setup.

Double wall minimum focal film distance is 91 cm.

Sphere or tubehead may be positioned to facilitate ease of exposure.

This angle determined by sphere diameter.

10° to 15°

Primary beam should be as near perpendicular as possible, but should not superimpose.

Equally spaced numbering not to exceed geometric radiation distortion principles.

All lead numbers should be placed at least 0.1 in. (0.3 cm) out of heat affected zone on all welds.

Block thickness and type equal to total material in beam path.

Material type

Material thickness

–2% of single wall

and block

Identification plate

Top view

Placement of image quality indicator and equivalent metal block

Film

Film in contact

Identification plate on top if cassette under sphere

Cassette and film

Top weld

Bottom weld

Radiography of Closed Tanks

Figure 6.26 shows some of the procedures for radiographing a closed tank when the X-ray tube or film cannot be placed inside. A single source is shown at various positions. The source position at one end of the tank illustrates that the other end of the tank can be covered with film and exposed with a single exposure. If the circumferential weld at the tank end should be on a cross sectional plane in relation to the source positioned at the tank end, additional exposures must be taken through the horizontal plane, as represented by the source position at the upper left. Geometric principles and minimum distortion distance must be maintained.

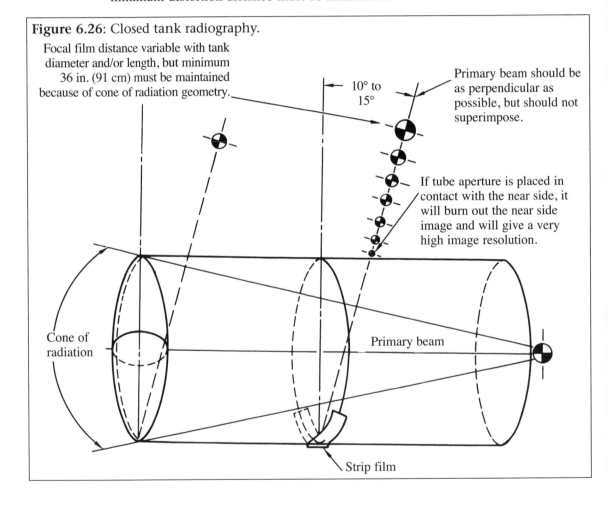

Figure 6.26: Closed tank radiography.

Focal film distance variable with tank diameter and/or length, but minimum 36 in. (91 cm) must be maintained because of cone of radiation geometry.

10° to 15°

Primary beam should be as perpendicular as possible, but should not superimpose.

If tube aperture is placed in contact with the near side, it will burn out the near side image and will give a very high image resolution.

Cone of radiation

Primary beam

Strip film

Radiographic Multiple Combination Application

Figure 6.27 illustrates a good method to use when setup is difficult, exposure time is excessive or material type and thickness are unknown. It is not recommended as a standard practice because density and sensitivity do not always measure to the required values through the various screens behind the first cassette. The back screens will filter rather than intensify. This application permits a high degree of latitude with a single exposure and may be used for weld grindouts where the depth cannot be checked or is unknown, or when the weld may have multiple grindouts of varying depths. This setup gives varying degrees of film density from the top film through the various films and screens to the back film.

Figure 6.27: Multiple combination application.

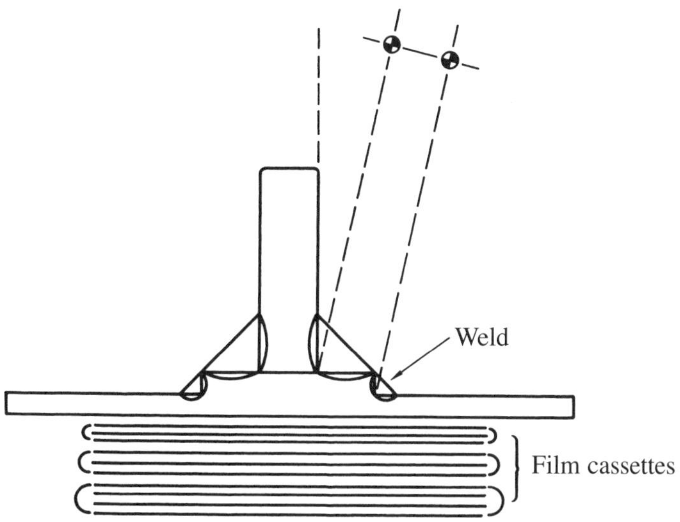

1. Vary voltage, amperage and time.
2. Vary distance (36 in. minimum).
3. Vary tube thickness and atomic number of tubehead filters.
4. Vary film types and combinations.
5. Vary screen and nonscreen combinations.
6. Vary number of cassettes under test object with the above combinations.

Radiography of Hemispherical Sections

All welds or seams on a hemispherical section may be radiographed with a radioisotope source, as shown in Figure 6.28. The source is placed in the geometric center of the section, and film is placed over all the welds. The gamma ray exposes all areas simultaneously. This procedure is time saving and is often used when gamma radiography is acceptable.

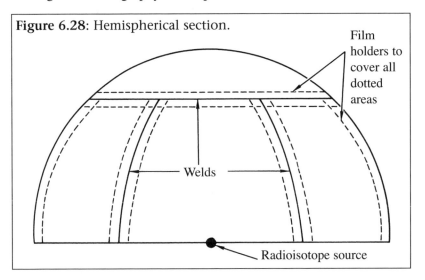

Figure 6.28: Hemispherical section.

Film holders to cover all dotted areas

Welds

Radioisotope source

Panoramic Radiography

Figures 6.29 and 6.30 illustrate two examples of panoramic exposure X-ray. Figure 6.29 depicts a means of radiographing welds on piping whose diameter is great enough to permit insertion of a rod anode X-ray tube. The beam of this type tube will expose the entire circumference of the pipe. The X-ray tube is placed in the center of the pipe so that the beam strikes the area of interest (the weld). Exposure calculations are based on the weld thickness. If gamma radiography is acceptable, a radioisotope source may be used in the same manner as a rod anode tube. The arrangement shown in Figure 6.30 is used when a sufficient number of similar small test objects are to be radiographed.

Figure 6.29: Panoramic radiography of a large pipe weld.

Source

All film holders exposed simultaneously

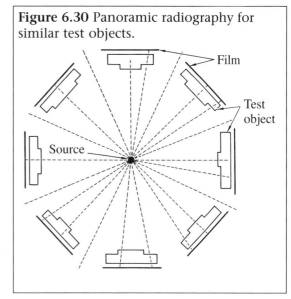

Figure 6.30 Panoramic radiography for similar test objects.

Film

Test object

Source

Radiography of Large Pipe Welds

Recommended radiographic procedures for large pipe welds that cannot be handled by elliptical or single wall shots are shown in Figure 6.31. Because it is impractical to obtain the desired results with a single exposure, the circumference of the weld is divided into three or more segments, and each segment is radiographed. The contact source is opposite to the film for three exposures. Exposure calculations are based on the thickness of the test object (double wall) penetrated in the area of interest of each segment.

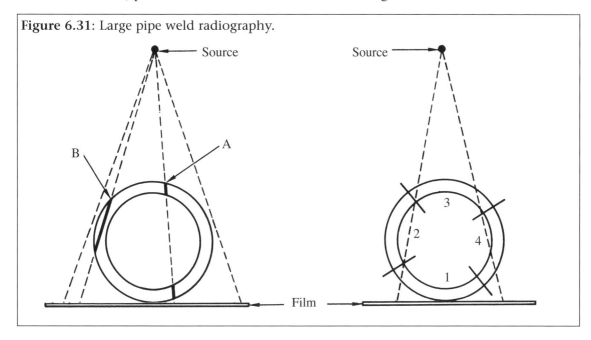

Figure 6.31: Large pipe weld radiography.

Source

B

A

Film

Source

3

2

4

1

Film

Radiographic Techniques of Discontinuity Location

Alignment

Figure 6.32 illustrates why discontinuities are often not recorded on the radiograph. Either the discontinuity cross sections are less than 2% of the overall test object thickness, or the longitudinal dimension of the discontinuity is not aligned with the radiation path. Figure 6.32b shows incorrect discontinuity alignment because the width of the discontinuity is less than 2% of the overall thickness. Figure 6.32c shows correct discontinuity alignment because the length of the discontinuity is more than 2% of the total thickness.

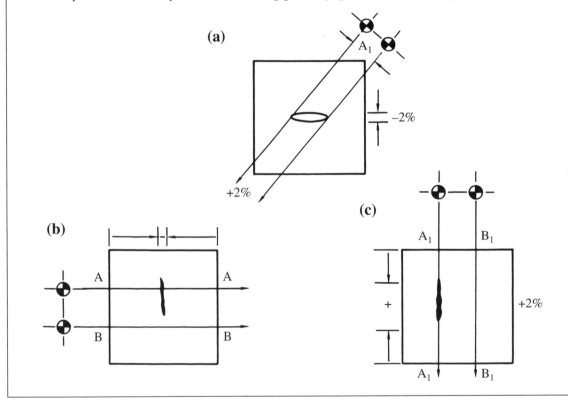

Figure 6.32: Angulation and discontinuity alignment. Fine discontinuities are detected more easily when the X-ray is directed along path A_1A_1 rather than along AA.

Discontinuity Depth Location Techniques

Superimposed single exposures may also be used for discontinuity location. The technique involves placing lead markers on both sides of the test object, exposing two separate films (each at a precise density) then laying one film over the other (superimposing the two back markers). The shift of the discontinuity is measured and calculated.

A simpler variation is to expose two separate films and superimpose the two back markers. If the shift of the discontinuity is less than one half the shift of the front marker, the discontinuity is nearer the film. If the shift of the discontinuity is greater than one half the shift of the front marker, the discontinuity is nearer the top or away from the film.

Radiography of Brazed Honeycomb

Figures 6.33 through 6.36 illustrate four types of exposures used to evaluate brazed or bonded honeycomb. Other special applications may be used; however, they normally will be variations of these four and are used for examination of a specific detail or area.

Double Surface Radiographs

Variations of this technique should be used to radiograph panels less than 1 in. (2.5 cm) thick. The following conditions should be satisfied on all exposures.

1. The upper surface fillet of any cell in the radiographed area should not overlap the extreme lower fillet of the adjacent cell.
2. The upper surface fillet of any cell in the radiographed area should not be superimposed on any other fillet.
3. The direction of the central beam of radiation should always be normal to the core ribbon direction, as shown in Figure 6.33.

Single Surface Radiographs

Variations of this application should be used to radiograph panels 1 in. (2.5 cm) or greater in thickness. The upper surface fillets (those closest to the X-ray tube) should be sufficiently blurred to permit adequate viewing of all lower surface fillets within the area radiographed.

A wedge shaped copper filter should be used at the X-ray tube, as illustrated in Figure 6.34, to obtain a more uniform density over the exposed area. Filter size and thickness should be adjusted for each X-ray tube.

Figure 6.33: Double surface radiography.

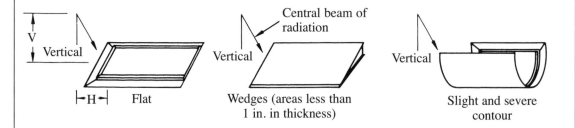

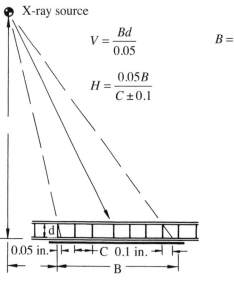

$$V = \frac{Bd}{0.05}$$

$$H = \frac{0.05B}{C \pm 0.1}$$

Flat and wedge panels

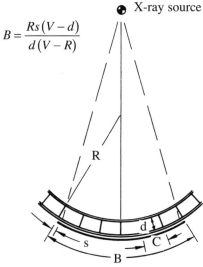

$$B = \frac{Rs(V - d)}{d(V - R)}$$

Contour panels

Legend

V = Perpendicular distance from tube to
  film plane
H = Horizontal distance from perpendicular
  to section of core being radiographed
B = Length of core section being radiographed

C = Cell size
d = Thickness of core section
R = Radius of contour of test object
s = Arc length

Edge Member Exposures
Two basic setups for edge member exposures are illustrated in Figure 6.35. Variations of these setups shall be used on all edge member exposures as outlined.

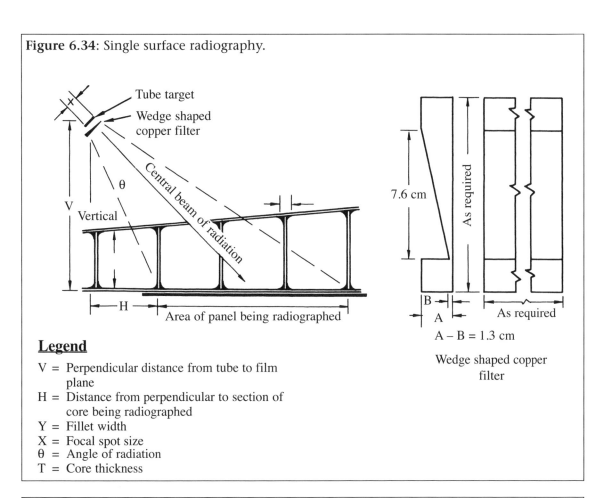

Figure 6.34: Single surface radiography.

Tube target

Wedge shaped copper filter

θ

Central beam of radiation

V

Vertical

H

Area of panel being radiographed

7.6 cm

As required

B

A

As required

A − B = 1.3 cm

Wedge shaped copper filter

Legend

V = Perpendicular distance from tube to film plane
H = Distance from perpendicular to section of core being radiographed
Y = Fillet width
X = Focal spot size
θ = Angle of radiation
T = Core thickness

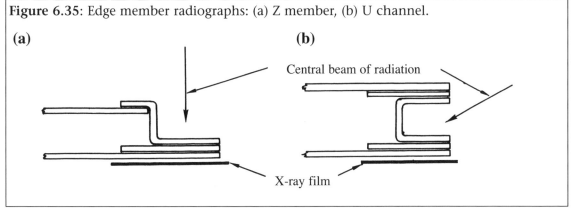

Figure 6.35: Edge member radiographs: (a) Z member, (b) U channel.

(a)

(b)

Central beam of radiation

X-ray film

Figure 6.35a illustrates the setup for Z edge member exposures. In most instances, adequate coverage will be obtained (on flat and slight contour panels) on core exposures.

Figure 6.35b illustrates the setup for wedge U channel (rib and spar) exposures. Both surfaces shall be radiographed separately. This permits determination of the amount of void area in both edge member surface.

Vertical Tie Exposure
Figure 6.36 illustrates the basic setups for vertical tie exposures. Variations of these setups should be used on all exposures made to evaluate the braze between the core vertical edge and the Z member vertical or the U channel vertical leg.

The setup for vertical braze evaluation (Z member) on contour panels and special exposures on flat panels is illustrated in Figure 6.36a. The vertical leg of the Z member should be inclined 8° to 10° from vertical. The central beam of radiation should be vertical and directly over the vertical leg of the Z.

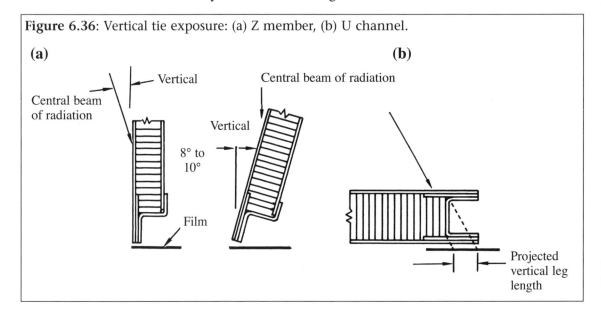

Figure 6.36: Vertical tie exposure: (a) Z member, (b) U channel.

An alternate setup would be with the vertical leg of the Z horizontal and parallel to the film. The central beam of radiation should be 8° to 10° off of vertical and directed toward the center of the area being radiographed.

The setup for vertical braze evaluation (U channels) on wedges and special exposures on all other panels is illustrated in Figure 6.36b. The lower surface (right hand section) should be the farthest from the X-ray tube and must be in the horizontal plane.

The central beam of radiation should be at an angle and distance so the projected vertical member height will not be less than half and not greater than the actual vertical leg height.

Radiography of Semiconductors

The application of radiography of semiconductors is somewhat different than applications discussed previously. With semiconductors, two major areas are of concern after the electrical acceptance tests have been completed. These are inconsistent internal construction and internal foreign material. Specific discontinuities associated with semiconductors are listed below (see Figure 6.37).

1. Loose particles, solder balls, flakes, weld splash and wire.
2. Loose or discontinuous connecting leads between internal elements and external terminals.
3. Extraneous matter, excessive solder or weld extrusions.
4. Inclusions or voids in seals or around lead connections or insufficient sealing material.
5. Inadequate clearance.

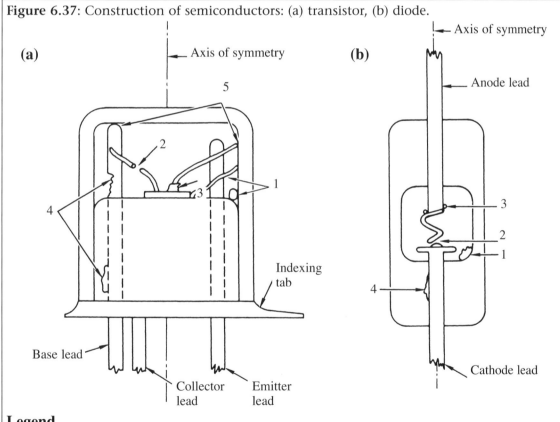

Figure 6.37: Construction of semiconductors: (a) transistor, (b) diode.

Legend

1. Loose particles, solder balls, flakes, weld splash, wire.
2. Loose or open leads between internal elements and external terminals.
3. Extraneous matter, excessive solder or weld extrusions.
4. Inclusions or voids in seals or around lead connections.
5. Inadequate clearance.

Techniques of Semiconductor Radiography

The following parameters must be taken into consideration to obtain satisfactory test results.

1. A beryllium tubehead or equivalent should be used.
2. Voltage must not exceed 150 kV; there is no limitation on current.
3. To avoid parallax, use extra fine grain, single coated emulsion film.
4. Use 20X magnification and sufficient light intensity during film interpretation to enable identification of 0.001 in. (0.003 cm) discontinuities.
5. Use correct semiconductor alignment.
6. Correctly locate radiographic source.
7. Ensure proper density in area of interest.

Alignment of Semiconductors

Figure 6.38 illustrates a typical holding fixture designed to curve the film to maintain equal source-to-film distance from the outer edge of the film to the center. The semiconductors should be mounted consistently, that is with the same pin on each facing the target.

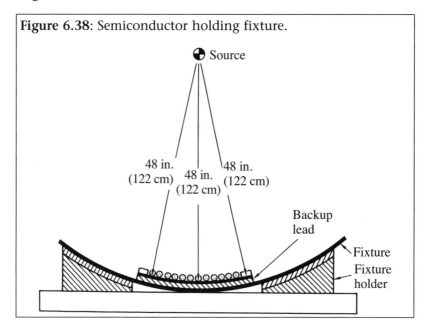

Figure 6.38: Semiconductor holding fixture.

Radiographic Views

Figure 6.39 illustrates the views required for satisfactory coverage of a transistor. Other views may be required to detect a specific type of discontinuity. Figure 6.40 illustrates the views required for satisfactory coverage of diodes, resistors and capacitors.

Figure 6.39: Suggested radiographic views of transistor.	Figure 6.40: Suggested radiographic views of diode, resistor and capacitor.

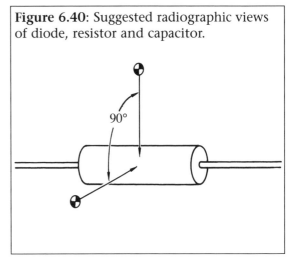

Fluoroscopic Application

Another method for examining transistors, diodes, resistors and capacitors is with a fluoroscope in conjunction with closed circuit television. Such a system would permit viewing the test object from different directions as it is rotated.

Chapter 7

Digital Radiographic Imaging

INTRODUCTION

This chapter presents approaches available to obtain digital radiographs by electronic means. The discussion and examples include techniques of conversion of X-rays to light and then to electronic images, photoconductive conversion of X-rays to electronic images, photostimulable phosphors, array detectors, line scan imaging and scanning electron beams.

Radioscopic digital imaging is related to radioscopy. In radioscopic imaging, the major emphasis is on the conversion of X-rays to analog electronic data that are viewed as video signals in real time. Digitization of these analog signals is a technique of digital imaging. Many of the principles for X-ray detection are identical, particularly where digital based cameras such as charge coupled device (CCD) cameras are used. This differs from radioscopic imaging in that the systems are not video based (although in some cases video could be the output). Rather, digital systems use discrete sensors with the data from each detection pixel being read out into a file structure to form the pixels of the digital image file.

An exception to the discrete sensor based systems discussed in this chapter is the photostimulable phosphor system that forms a latent image (similar to film) on a storage phosphor imaging plate. The screen is read electronically using a special laser scanner. The pixelization in this case is based not on the X-ray sensitive phosphor but in the laser scanner process.

Development

The ability to develop digital imaging technology that would be useful for radiographic testing is due in large part to the growth in the speed and memory of computer systems. In the 1980s, images of 512×512 pixels with 8 bits of data (256 kilobytes) were considered large and created storage and display problems for the computer systems at that time. By the twenty-first century, image files of 1500×2000 with 16 bits of data (6 Mb) are common and can be transported, stored and displayed with relatively inexpensive computer systems.

The medical community has led the development of digital X-ray imaging. Spinoff from the medical systems has occurred, allowing the introduction of digital imaging technology for industrial radiography uses.

In the early 1980s, digital imaging for radiographic purposes was primarily done by electronic digitization of the video signal from a radioscopic system. Charge coupled device cameras were available, but the most common application was as a video output camera. Developments in direct digital image output for these cameras resulted in CCD arrays in the 1990s that consisted of millions of pixels.

Also developed in the 1970s and 1980s were digital imaging systems using line scan detector arrays. To form the image, either the test object or the detector array was physically scanned in the dimension perpendicular to the array. In the late 1970s to early 1980s, the photostimulable phosphor array was developed for medical use and was used in industry in the 1990s.

In the 1990s, the development of large thin film transistor arrays provided the tool that could make possible large area X-ray imagers using either amorphous silicon or amorphous selenium panels.

Detectors for Digital Imaging

Digital radioscopic detectors are used in numerous industries, from airport baggage scanning to medical diagnosis. In addition to these widely used applications, digital radioscopy is finding an increasing role for inservice nondestructive testing, as a diagnostic tool in the manufacturing process, for online production line testing and with conveyer handling systems. Digital radioscopic detectors are also being used as handheld devices for pipeline tests, as film replacement devices, in industrial and medical computed tomography systems and as part of large robotic scanning systems for coverage of large structures.

The digital image by its nature will provide numerical results important for metrology and thickness measurements. The development of a wide range of digital X-ray imaging products provides digital image data and results that can be incorporated into the massive digital manufacturing and services databases that have emerged to help manage the life cycles of products and structures.

In the field of industrial digital radioscopic, there is really no single standard X-ray system to address all applications. Economics, speed, quality and the impact on the overall manufacturing or service processes are key in designing and building digital radiographic systems. An important aspect of that design is the consideration of the digital X-ray detection device itself. For this selection, there are almost as many choices of detectors as there are ways to configure the overall test system. The different digital detector technologies available are discussed below.

Principles of Digital X-Ray Detectors

The detection devices that support the larger imaging systems are the following.

1. Phosphors deposited on amorphous silicon thin film transistor diodes.
2. Photoconductors such as amorphous selenium deposited on thin film transistors.
3. Phosphors deposited or coupled through fiber optic lenses onto charge coupled device detectors and complementary metal oxide silicon detectors.
4. Photostimulable storage phosphors.
5. Phosphors deposited on linear array systems
6. X-ray scanning source reversed geometry detectors.

Each of these devices has an X-ray capture material as its primary means for detecting X-rays. This material is either an X-ray phosphor material combined with a photoelectric device (diode, photomultiplier tube or charge coupled device) or is an X-ray photoconductor material that is then followed by an electronic readout device. The most common detection systems in operation today are the flat panel detection systems based on amorphous silicon and amorphous selenium structures, the camera systems based on charge coupled devices technology and the storage phosphor systems.

Each of these devices can be used to replace film radiographic techniques depending on the size of the application and on the spatial resolution, image contrast and speed required. The detectors have variable modes of operation or are available in different architectures to address diverse applications. There are numerous pixel architectures of amorphous silicon detectors, but it is important to note that currently not all detector choices allow real time operation of 30 frames per second.

Charge Coupled Devices

Scientific charge coupled devices, although they are typically small in size, have been made with high pixel densities. The fields of photography, astronomy and microscopy have demanded this, and the nondestructive testing industry has been a beneficiary of these developments.

Charge coupled devices (or CCDs) have not been fabricated into larger arrays because the CCD is based on crystalline silicon, which has traditionally been cut from silicon wafers available in sizes only as large as 4 to 6 in. (10 to 15 cm) in diameter or less. A larger field of view can be accomplished with CCDs through tiling of the devices or through a lens or a fiber optic transfer device to an X-ray conversion (phosphor) screen. The downside of the lens approach is that it has very poor light collection efficiency. Fiber optics or tiling do not provide large fields of view but will result in more efficient light collection.

Thin Film Transistor

Larger amorphous silicon and amorphous selenium detectors based on thin film transistor technology have been made commercially available with a pixel pitch smaller than 75 μm. Amorphous silicon through large area amorphous silicon deposition and processing/etching techniques offers a solution to the size constraints of CCDs while maintaining good light collection efficiency from the phosphor or photoconductor (selenium) material. Because the phosphor layer is typically deposited directly onto the silicon, efficient light transfer is easily obtained. However, the readout circuitry (described elsewhere) in these devices requires a large pixel space to accommodate the thin film transistor (TFT) and data lines and scan (gate) lines required for operation, thus limiting how small a pixel this device can permit.

Light Collection Technology

The amorphous silicon thin film transistor circuitry has a fill factor of active photodiode ranging from 65 to 90%. Charge coupled devices use a transparent polysilicon gate structure for reading out the device and have a fill factor of close to 100%. On a per pixel basis, the CCD is therefore more efficient in collecting the light produced from the phosphor material. For small field of view applications, the directly coupled CCD approach provides high spatial resolution and high light collection efficiency. For large field of view applications, the amorphous silicon approach offers excellent light collection efficiency (no lenses), in a thin, compact, sturdy package.

Radiation Conversion Material

The amorphous selenium device is similar to the amorphous silicon based detector. They both use thin film transistor readout circuitry. The difference lies in the X-ray conversion material. The amorphous selenium detector relies on the selenium photoconductive material (not a phosphor layer) as a means to detect X-rays. The selenium converts X-rays to electron hole pairs that then get separated by the internal bias of the device and captured by an electrode structure. The amorphous silicon thin film transistor circuitry beneath the selenium layer provides readout of the charge with the aid of field effect transistors (FETs) in a similar manner to that of the amorphous silicon detectors. The selenium layer is typically 0.02 in. (0.05 cm) thick.

For applications with large fields of view, amorphous selenium offers direct X-ray collection efficiency in a sturdy, compact package.

Storage Phosphors

Storage phosphors trap X-ray induced charge carriers in the color centers of such phosphor materials as europium activated barium fluorobromide (BaFBr:Eu). Although prompt phosphorescence occurs during X-ray exposure, some of the charge trapped in the phosphor material is stored in the color centers in the crystalline structure. The stored charges can be released when stimulated by infrared or red laser light. The rerelease of trapped charges subsequently creates photostimulated luminescence of the same emission wavelength that the earlier emission process produced.

A photomultiplier tube converts the emitted photostimulated luminescence to an electrical signal that is then amplified and sampled. These systems have a practical spatial resolution and contrast sensitivity and have been widely used in production radiography. Additionally, they are used like film and are somewhat flexible. They are also portable in the field and fully reusable.

These screens have to be transferred to a laser processor before they can be interpreted. This removal process is where this technology departs from the other digital approaches. Photostimulable luminescence techniques can be more productive when imaging plates can be used in the field in a collection or batch that covers large areas for each exposure.

The main advantage of phosphor screens over film is the reduction of costly film use, the ability to digitally acquire a film quality image, the dynamic range and the corresponding benefits of that digital image file, such as easy archival and retrieval.

Linear Arrays

Linear array detectors are much like CCDs, except they typically only have pixels in one dimension or they may be composed of a small rectangular array such as a 32×1024 pixel array. The advantage of linear arrays is their scatter rejection capability. Scatter radiation exiting a test object can be a large contribution to the degradation of the contrast in the image. The linear array system acquires its image by being scanned one line (or a group of lines) at a time across an object. The key is that the radiation beam is masked or collimated to match the size of the detector. This dramatically decreases the object's scatter field. The scatter detected at each of those lines is substantially less than that of individual lines in an area array. Linear arrays have been successfully used in computed tomography applications and have also been found to be effective for digital radiographs.

Scanning Beam, Reversed Geometry

The reversed geometry system goes one step further in reducing X-ray scatter in the test. In this case, the data are acquired with a small thallium activated sodium iodide (NaI:Tl) scintillator coupled to a photomultiplier tube. A large scanned X-ray source with a target diameter of about 10 in. (25 cm) is used to define the image. The

X-ray source operates in a manner similar to a video monitor. An electron beam is electronically rastered over the inner surface of the front of the X-ray source. Where the electrons collide with the inner surface of the tube, X-rays are generated. By electronically scanning the electron beam, the instantaneous position of the X-ray source is scanned over an area of the front surface of the tube. The size and location of the scanned region is user definable, variable from 0.25 to 16 s. The acceleration voltage is also user definable from 55 to 160 kV with an electron beam current up to about 0.5 mA. The diameter of the electron beam spot at the inner surface of the tube is about 0.001 in. (0.003 cm).

The test object is placed on top of the X-ray source. This is the opposite of conventional radiography where the object is placed near the imaging detector and the source is a point source. The data acquisition computer also controls the rastering of the electron beam. By acquiring the output of the detector as a function of electron beam position, the computer can generate a real time radiograph of the test object.

Because a single small area detector is used and the object is placed at the source, not at the detector, the X-ray scatter from the object is effectively zero. The disadvantage of this approach is that, because it is reversed geometry, the effective focal spot size is that of the detector size. The detector size is typically much larger than a typical industrial X-ray focal spot. Any test object that has some thickness will show significant unsharpness as the feature of interest moves away from the X-ray source.

Detection Efficiency

With the exception of the photoconductive selenium based detector, all detectors given here use a phosphor layer of one sort or another to capture and convert the X-ray intensity. The selection of the phosphor or photoconductive material, its thickness and effective atomic number will affect the total number of X-rays absorbed in the conversion material. Once energy is absorbed each material, phosphor or photoconductor, has its own efficiencies for conversion of this energy into either light or charge carriers. There are other coupling steps following this to transfer the signal onto the pixelized readout circuitry. The performance of the X-ray detector to convey the information in the radiation beam is then dependent on the efficiency of each step in the X-ray conversion process leading to an electronic signal. The signal-to-noise ratio of the detector and thus the image contrast are dependent on the transfer of information along the imaging chain.

SPATIAL RESOLUTION

The efficiency of the energy conversion process relates to the speed of the examination, the throughput and the tradeoff with contrast sensitivity (the ability to detect a small change in thickness or density).

Detector Resolution

The spatial resolution of the detector determines if features in the object are detectable from a pixel sampling consideration. The selection of the spatial resolution of the detector is also important in designing or selecting a detection system. From the aspect of image contrast and spatial resolution, it is desirable to have the largest pixel that will allow detection of the features of interest in the radioscopic examination. For example, it is not necessary to select a 39 μm pixel pitch if the application is for the detection of large foreign objects left behind in an engine. Similarly, fatigue crack detection is probably not going to be too successful with a pixel pitch of 200 μm or larger.

Pixel Pitch

The predominant factor that governs the spatial resolution of a detector is the pixel pitch. The selection of the radiation conversion screen then becomes important. Here the architecture of the radiation conversion material will dictate to what degree the full spatial resolution of the detector can be realized.

As the pixel pitch is reduced to increase resolution, the total number of pixels in the image increases for a constant field of view. The file sizes for typical images run from 2 to 8 Mb. However, for digital images at radioscopic (real time) frame rates of 30 frames per second, the image size must be closer to about 1 Mb at current technology. Therefore tradeoffs are made in selecting larger pixels for smaller fields of view.

The selection of a high atomic number radiation conversion material that can provide a signal gain sufficient to not allow secondary quantum sinks following absorption is critical. Forming this material into a shape that directs the signal onto a single pixel, as is done with cesium iodide, is then crucial to maintaining good image detail. As the atomic number of the conversion material goes down, the percentage of X-ray information in the radiation beam will diminish and poor image contrast will result.

To compensate for this image degradation, the conversion material can be made thicker or, if time permits, a longer total exposure time can be selected to capture more X-rays. Making the conversion layer thicker can affect the spatial resolving power of the device because both X-ray cross talk and signal (light or electron hole carriers) cross talk will yield a breakdown in the modulation of the signal somewhere in the spatial frequency range of the detector.

Modulation Transfer Function

A good measure of the spatial resolution is the modulation transfer function (MTF). The modulation transfer function measures the signal modulation as a function of spatial frequency and is typically computed using a fourier transform of a line spread function acquired on an angled tungsten edge placed directly on the detector.

Figure 7.1 shows the power of a modulation transfer function for revealing a breakdown in spatial resolution throughout the spatial frequency regime of the detector. If the spatial resolution drops near 0 line pairs per millimeter, this drop can be interpreted as a severe degradation in image contrast and will result in poor density discrimination. If the modulation transfer function is low at high spatial frequencies, near the sampling limit of the detector, this indicates that the conversion material is not a good choice for detection of the fine features that the system was designed to detect. Another choice should be selected. Balancing the spatial resolving powers of a conversion material with its quantum efficiency has been an active area of research and development in digital radioscopy since 1985.

Figure 7.1: Localized variations in modulation transfer function curves in some regions of the spatial frequency domain. Maximum spatial resolution for the detector is 10 line pairs per millimeter.

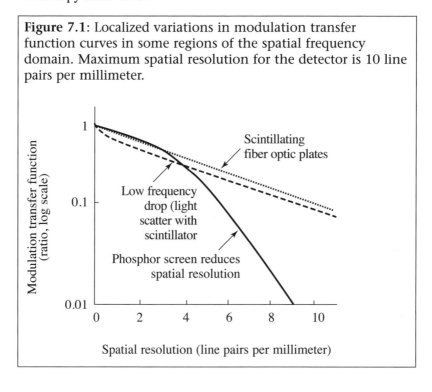

Gain and Offset Correction

Imagery from digital detectors is frequently normalized for pixel-to-pixel gain variations and also adjusted to subtract out the background or offset. The offset or background signal is usually a small percentage of the maximum signal and is common to all digital detectors. It is important to subtract this background signal to provide a wider linear range and to subtract any latent images on the detector. In performing a gain correction, not only are pixel-to-pixel variations reduced but also variations in the optical components feeding these pixels will be diminished. Performing this gain correction can also be used to flatten the radiation intensity distribution across the detector panel. Making the radiation beam intensity more uniform across the detector can result in wider latitude (viewable thickness range) in the image. This normalization is really not possible with film radiography.

The gain correction is accomplished by taking an image with a radiation technique similar to that planned for production but without an object in the beam (an air image) and with a much reduced radiation intensity. By simply performing an image division by the gain factor on a pixel-by-pixel basis, the offset corrected air image is then used to correct each subsequent image of an object. Following gain and offset correction, detection sensitivity improves in relation to an image that does not have this correction. For the air image, it is critical that the image be free of transient latent images, have the correct intensity and also not contain an object of any sort (such as a fixture) in the beam. If any of these occur, then every subsequent corrected object image will contain artifacts and the correction will do more damage than good.

Radiation Damage

In digital imaging devices, there are numerous elements of the detector assembly that can be damaged by the ionizing radiation. Every component in the imaging chain not shielded appropriately from X-rays or gamma rays can be damaged. The term *radiation damage* is a general term that can refer to any range of damage to a component in the detection chain. The damage can lead to subtle changes in performance, all the way to failure.

Most digital detectors are designed so that the electronic components behind the radiation conversion material are either shielded from the radiation (for example, by the conversion material itself or by fiber optic transfer components behind it) or are sufficiently thin to absorb only a small portion of the radiation that impinges on the component. The damage that occurs in the electronic circuitry can result in an increase in the electronic noise of the device and eventually to failure as the accumulated dose to the component increases. Each manufacturer uses proprietary circuitry and various forms of shielding elements to prevent these effects.

The radiation conversion material, being the primary radiation absorption component, is exposed to the highest levels of radiation within the imaging chain. Phosphors such as cesium iodide and photoconductive materials such as selenium have areas within their band structures that will trap electron and hole carriers produced by the ionizing radiation. In many circumstances, thermally released carriers from these traps will yield a delayed luminescence or a delayed release of charge. This form of radiation damage known as *afterglow* or *lag* usually increases as a function of radiation dose until an equilibrium occurs where the number of carriers being trapped equals the number being thermally released.

Another form of radiation damage to radiation conversion materials is when the carriers are permanently trapped in deep centers within the band gap. This trapping is sometimes associated with a darkening of the conversion material and usually results in a rapid decrease in signal that can only be healed by heat annealing of the material or by slow thermal release at room temperature. This form of damage is known as *gain decrease*. In other materials, it is possible to observe a rapid signal gain increase as a function of increased radiation dose. Although the mechanism of gain decrease is not widely understood, both gain changes can impart spatial artifacts into a current image created by the variation in radiation intensity across a prior test object image. In most cases, these gain changes are not long term or permanent. If the system is prone to these radiation induced gain changes, it is important to continually update gain and offset data, even if the actual examination is not changing, so that these artifacts can be reduced. If the problem becomes severe it might warrant a new phosphor.

Selection of Systems to Match Application

Some of the key characteristics that might be considered in the selection of a digital radioscopic imaging system are the following.

1. Detection precision and accuracy.
2. System speed to match that of manufacturing and examination processes.
3. Area of the detector to match manufacturing throughput.
4. Volume of the device for access to tight locations in an assembly.
5. Presence of artifacts that can impact detection capability.

If a large area detector is needed and there is a requirement to work at real time frame rates of 30 frames per second, then a CCD detector should be selected. If static imaging is required but the highest spatial resolution is needed and the object size is not large, then a system using a low noise phosphor or CCD should be selected. For this same application, a large area flat panel detector operating in static mode can also be selected if used in combination with a microfocus X-ray tube but only if the application can withstand the longer exposure times associated with magnification radioscopy.

If super high resolution is required, for example, very tight small crack detection, then magnification may be required with the high resolution CCDs.

It is important to have the largest pixel that can be accepted from a feature detection (spatial resolution) standpoint. This parameter then provides the highest throughput possible because larger pixels can produce a higher signal-to-noise ratio for a given X-ray exposure. Larger pixels will also allow a lower exposure for a constant signal-to-noise ratio. Larger pixels permit thicker X-ray conversion materials, again potentially adding speed to the examination. Finally, larger pixels will result in a larger overall field of view (larger throughput). For example, a four million pixel array of 200 μm pixels will have 16 times the field of view of a four million pixel array of 50 μm pixels.

As mentioned earlier, the size of the detector and the size of the pixel still go hand in hand using today's technology. It is possible to have a 10 000 × 10 000 pixel array of 25 μm pixels resulting in a 10 × 10 in. (25 × 25 cm) device. That said, if a smaller pixel device is selected, it might be possible to average pixels into larger superpixels to enhance speed and part throughput. The minor drawbacks of such superpixels is that the X-ray conversion material may not be of optimal thickness for the larger size pixel, and the percent of active pixel (because the amorphous silicon approach may be summing four field effect transistors) may not be as great as if the pixel were designed with a single set of readout circuitry. Finally, the noise of averaging four pixels is a little higher than the noise of a similar detector element of the same size.

For tight locations, small detectors based on CCD or complementary metal oxide silicon technology can be used.

Where the requirement is to simply replace film in favor of a lower cost digital solution, then storage phosphors can be used quite successfully. However, if access is not an issue, then the other digital approaches may be more cost effective over the long term because they are more amenable to high speed mechanized automation of the detector and X-ray tube to scan a test object or conversely for the test object to be scanned through the stationary tube and detector configuration.

Linear arrays can be used in an assembly line configuration, as can the real time flat panel and CCD detector systems. Line scanners offer the advantage of reduced sensitivity to X-ray scatter in relation to area array systems.

The scanning beam, reversed geometry system has shown promise in reduced access applications because this detector module is quite small. The reversed geometry system is probably the best system for reduced sensitivity to X-ray scatter because the detector is essentially a point based sensor. However, it is important to note that the detector is typically much larger than an X-ray tube focal spot. Because of the effective focal spot size of the system, there may be some geometric constraints placed on this system in terms of image unsharpness.

Artifacts have been prevalent in digital radiographic systems. The presence of artifacts, therefore, has to be evaluated almost on a detector-to-detector basis.

X-Ray Detector Technology

Amorphous Silicon Detectors

Most new amorphous silicon designs are based on a flat glass panel that has undergone a deposition process resulting in a coating on one side that contains several million amorphous silicon transistors. These transistors are arranged in a precise array of rows and columns. Bias and control lines are brought to the edge of the panel for each individual transistor. The length and makeup of these control lines play a role in how fast image data can be scanned out of the array. On large receptors the control lines are typically brought out from the middle to both sides of the panel to minimize the track lengths.

Figure 7.2 illustrates a cutaway view of a typical panel design. This configuration is typical of a receptor incorporating a phosphor conversion layer. The phosphor layer converts the X-ray photons to light photons. The light photons are in turn converted to electrons by the amorphous silicon array and the readout electronics.

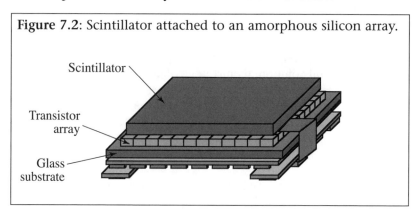

Figure 7.2: Scintillator attached to an amorphous silicon array.

The next layer of the assembly (shown in Figure 7.2) is the amorphous silicon transistor array. Deposited on a glass substrate to provide a rigid and very flat surface, this layer converts the light photons, from the phosphor, into electrons that can be read out, amplified, digitized and stored as an image. Each element of the amorphous silicon array is made up of a transistor and a photo diode. See Figure 7.3 for a schematic representation of a small section of the receptor. The light from the phosphor is captured by the photodiode and then read out through the transistor in a very high speed and synchronized process. Each charge is digitized by an analog-to-digital converter and then stored in a precise memory location in the image processing computer. Once every transistor is sampled and read out, a complete image will be displayed on the viewing monitor.

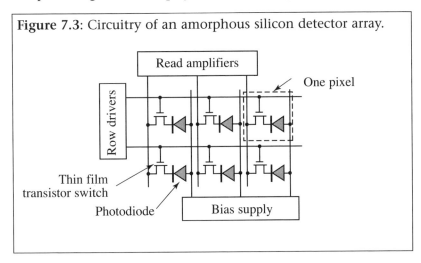

Figure 7.3: Circuitry of an amorphous silicon detector array.

With regard to flat panel receptors, a pixel is the area of one transistor and one photodiode. Typically these pixels range is size from 0.004 × 0.004 in. (0.01 × 0.01 cm) to 0.002× 0.002 in. (0.004 × 0.004 cm). A typical pixel is shown in Figure 7.4.

Figure 7.5 shows a cross sectional view of an amorphous silicon receptor that uses a cesium iodide scintillator. This view suggests the path taken by the X-ray beam as it exits the object being examined and enters the input of the receptor. Figure 7.6 shows an image receptor assembly with the electronics folded out from behind the panel during assembly and testing. The large dark area at the right center of the picture is the amorphous silicon array. Around the edges of the array are all the electronics required to control and read out the image data.

The photograph in Figure 7.6 shows the mechanical makeup of the amorphous silicon layer. The bias and data lines provide the ability to properly control each pixel and the connections used to get the image data out.

The high resolution view of an aluminum tube weld in Figure 7.7 was obtained with an amorphous silicon panel originally acquired at four times geometric magnification with a microfocus X-ray source. In the lower portion of the image is the placement of a 0.003 in. (0.007 cm) thick ASTM aluminum plaque image quality indicator. The central hole in this plaque is 0.01 in. (0.03 cm) in diameter and is clearly visible through the thin aluminum walls, about 0.05 in. (0.13 cm) thick. This image illustrates that very small diameter porosity can be detected in these structures: comparing the gray scale and size of the pores in the weld with the holes of the image

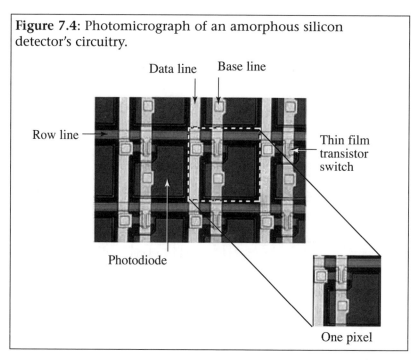

Figure 7.4: Photomicrograph of an amorphous silicon detector's circuitry.

Data line

Base line

Row line

Thin film transistor switch

Photodiode

One pixel

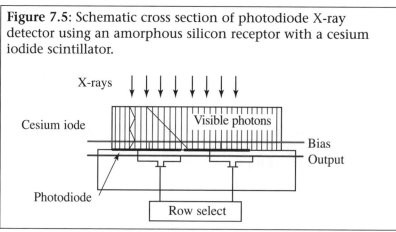

Figure 7.5: Schematic cross section of photodiode X-ray detector using an amorphous silicon receptor with a cesium iodide scintillator.

X-rays

Cesium iode

Visible photons

Bias

Output

Photodiode

Row select

Figure 7.6: Photograph of an amorphous silicon detector with electronics mounted to the side of the panel in a position where they can be shielded from X-rays.

quality indicator reveals porosity much smaller in diameter than the 0.01 in. (0.03 cm) hole.

Figure 7.7b provides a high pass filter rendition of this image. Once filtered, contrast may be added to the image so that high contrast can be observed across the entire thickness range of the object. This now provides information on the weld almost to the tangent point and assists the operator in identifying discontinuities over a wider range of thickness in a single view.

Most flat panel receptors are designed to provide radiographic acquisition capability at a rate of one image about every 5 to 10 s. Some designs take more or less time to put the image on the monitor but most fall into this range. This speed is much faster than what can be achieved with film cassettes.

Figure 7.7: Aluminum tube weld image acquired with an amorphous silicon detector with four times geometric magnification: (a) porosity as small as 0.005 in. (0.013 cm) can be detected in gray scale image; (b) high pass filter provides high contrast over wider thickness range in single view, making porosity evident almost to the tangent point of the weld.

(a) 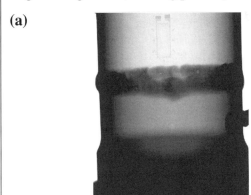 **(b)**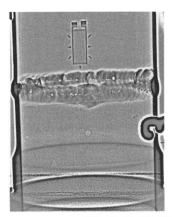

Amorphous Selenium Detectors

In flat panel arrays using an amorphous selenium converter (or other photoconductors), the X-ray to electrical charge conversion process is referred to as *direct conversion* because no intermediate steps are required. The high voltage bias field applied to an amorphous selenium layer creates vertical field lines. Because the field lines are parallel to the incident X-ray beam (other than for oblique angles), the field prevents the charge from lateral scattering and thus there is virtually no blur. Intuitively, this would seem to suggest that the amorphous selenium conversion layer (excluding the pixel electrodes) should exhibit extremely high resolution. In fact, measurements prove this to be the case.

Charge Coupled Device Radiographic Systems

Charge coupled devices (CCDs) are used in X-ray imaging systems in combination with X-ray phosphors or scintillators without the need for electronic image intensification. A CCD is an integrated circuit formed by depositing a series of electrodes, called *gates* on a semiconductor substrate to form an array of metal oxide semiconductor (MOS) capacitors. By applying voltages to the gates, the material below is depleted to form charge storage wells. These wells store charge that is injected into the CCD or generated within the semiconductor by photoelectric absorption of optical quanta. If the voltages over adjacent gates are varied appropriately, the charge can be transferred from well to well under the gates, much in the way that boats will move through sets of locks as the potential (water heights) are adjusted.

In the simplest design, the CCD is rapidly scanned to provide television frame rates with typical exposures per frame of 33 ms. In this mode, the signal can be very low and the resulting signal-to-noise ratio will therefore also be low because of the small number of photons impinging on the phosphor in the time allotted and the high noise level of the CCD. The noise of the device increases as a function of the square root of the readout speed and is quite high at real time frame rates. The image quality can be improved by averaging multiple frames in a digital processor but the high noise of the device operating at these speeds does not provide film quality images.

The better way to improve the signal-to-noise ratio using CCDs, is to integrate the charge produced by light from the phosphor directly on the CCD cells. The wells generated by the readout approach can be sufficiently deep to capture three to four orders of magnitude in equivalent light levels. Because the exposure times are now much slower than the real time rates of traditional CCD video cameras, the readout speed can be reduced to obtain lower camera noise levels. On a frame-by-frame basis, the signal levels have been increased while the additive noise from the camera has been decreased. In this mode, further electrostatic image intensification is not needed.

Charge coupled devices are available with image formats as large as 4096×4096 pixels and 16 bits. Some devices have been made as large as 2.4×2.4 in. (6.1×6.1 cm). A phosphor screen can be coupled directly to the CCD itself but even if the phosphor has good X-ray quantum efficiency, those X-ray photons not absorbed by the phosphor (even if it is a small percentage) can still be absorbed in the silicon layer of the CCD and yield a significant direct excitation speckle noise in the image. To avoid this noise, a fiber optic image transfer plate or a scintillating fiber optic plate may be used to absorb the transmitted X-rays before being absorbed in the silicon.

The field of view of CCD X-ray systems can be expanded with a fiber optic taper or a lens system. The results of these configurations are shown in Figure 7.8.

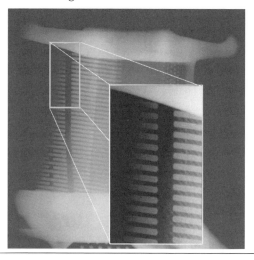

Figure 7.8: Radioscopic image of nickel alloy bucket blade with enlarged view of finning effect.

Fiber optic tapers are fiber optic face plates in which the size of each fiber in the face plate is reduced so that an image deposited at the input surface may be transferred to a smaller device such as a CCD at the output surface. Fiber optic tapers thereby increase the field of view, provide efficient light collection (with respect to a lens), offer shielding of the CCD from direct X-ray hits and can yield a compact, lightweight, rigid design. Fiber optic tapers have been incorporated with a 4 × 4 in. (10 × 10 cm) active area.

A lens as an optical coupling device has the drawback that it is a very inefficient light collection device. Compared to a fiber optic taper, a lens system is less efficient by a factor of ten or more. This inefficiency can lead to secondary quantum sinks and additional noise in the image. Secondly, the lens does not provide adequate shielding to the CCD, so an additional shielding glass is needed directly in front of the CCD to reduce direct X-ray hits on the device. In addition, a mirror can be used to move the camera out of the radiation beam. The CCD can then be shrouded in lead to reduce excitation by tangentially scattered X-rays. One advantage of a lens is the increased flexibility it offers to adjust the field of view for examination of both large and small objects.

Linear Detector Arrays

The linear detector array systems are ideally suited for production environments. Many industries — including automotive manufacture, cargo transport, food inspection, munitions, security and nuclear waste containment — use linear arrays of X-ray detectors for their testing needs. Thousands of these units have been installed. Figure 7.9 shows images of various objects suspended in barrels and detected with linear arrays.

Figure 7.9: X-radioscopic images of barrels and contents made with linear detector arrays.

Chapter 8

Special Radiographic Techniques

INTRODUCTION

Radiography is defined as a method using the penetration and differential absorption characteristics of X-rays and gamma rays to test materials for internal discontinuities. In this chapter, special radiographic techniques will be discussed.

FLUOROSCOPY

Fluoroscopy is the process in which an X-ray image is observed on a fluorescent screen, as seen in Figure 8.1. It is a relatively low cost, high speed process and is easily adapted to production line requirements. Its disadvantages include the following.

1. Cannot be used with test objects that are thick or of dense material because the intensity of the radiation passing through the test objects would be too low to sufficiently brighten the screen.
2. Relatively poor sensitivity because of the short source-to-screen distance required to obtain sufficient luminance, and the low contrast and coarse grain of the screen.
3. Does not produce a permanent record.

Despite these disadvantages, fluoroscopy is widely used in applications where rapid scanning of test objects for gross internal discontinuities or abnormal conditions is desirable. Using fluoroscopy, a number of test objects can be screened before submitting the lot to other radiographic tests and those with gross discontinuities can be immediately rejected with resultant cost savings.

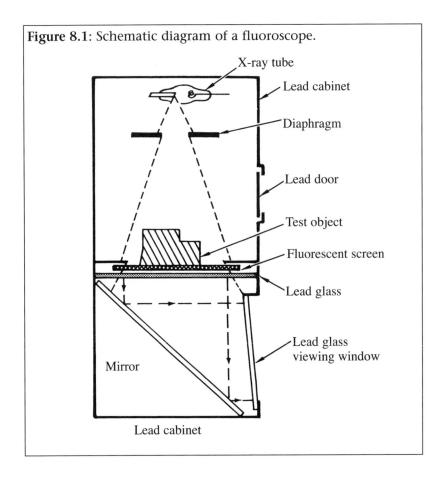

Figure 8.1: Schematic diagram of a fluoroscope.

X-ray tube

Lead cabinet

Diaphragm

Lead door

Test object

Fluorescent screen

Lead glass

Lead glass viewing window

Mirror

Lead cabinet

IMAGE INTENSIFIER

The image amplifier is designed to overcome the disadvantages of fluoroscopy and the relatively low luminance of its image. It also serves to protect the technician from radiation. It consists of an image tube and an optical system, as shown in Figure 8.2.

The image tube converts the X-ray image on the fluorescent screen to electrons and accelerates and electrostatically focuses the electrons to produce the image on the smaller fluorescent screen. The optical system magnifies the image on the small screen, and it appears as if the viewer was looking directly at a normal sized screen.

The luminance amplification factor is the product of the reduction in screen area and the energy of electron acceleration. Dependent on image tube design and construction, this factor ranges from 100 to 1000. Using a suitable camera and a closed circuit television system, the X-ray image produced by the image amplifier may be viewed on a monitor screen. Or, if desired, the image may be photographed or videotaped to produce a permanent record .

Figure 8.2: Schematic of an image amplifier.

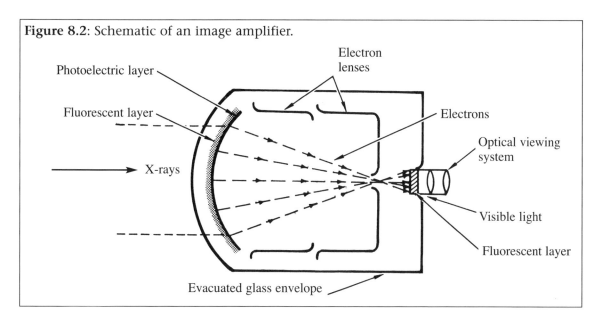

TELEVISION RADIOGRAPHY

The television technique is relatively inefficient because there is a large energy loss incurred in converting the X-rays into light, which is in turn converted into electrical signals that energize the television system. Advanced techniques are available that use television equipment specifically designed for radiographic applications.

The X-ray sensitive vidicon tube, shown in Figure 8.3, is an example of this type of equipment. The tube differs from normal vidicon tubes in that it is X-ray sensitive rather than photosensitive. It is widely used to permit instant image reproduction, combined with observer protection from exposure. The tube is the key component of a system otherwise consisting of an X-ray source that provides an intense small diameter beam, a unit for handling and positioning test objects and a closed circuit television readout. The system is designed for radiographic testing of small test objects such as electronic assemblies. It is also suitable for in-motion X-ray testing. Permanent records may be obtained by photographing the monitor screen of the readout system.

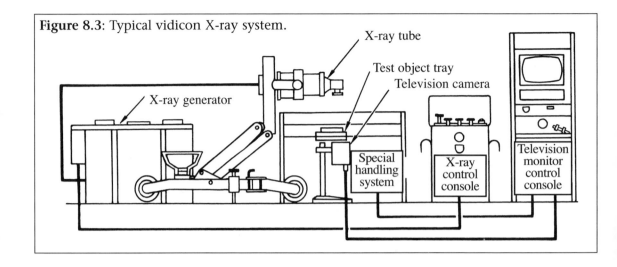

Figure 8.3: Typical vidicon X-ray system.

X-ray tube

Test object tray

Television camera

X-ray generator

Special handling system

X-ray control console

Television monitor control console

Xeroradiography

Xeroradiography is a dry radiographic process that uses electrostatically charged plates to record an X-ray image. The basis of the process is the peculiar characteristic of selenium that causes it to become a relatively good electrical conductor when exposed to X-rays.

The plate used to record the X-ray image consists of a thin layer of selenium bonded to a backing plate of aluminum. Under darkroom conditions, an electrostatic charge is placed on the selenium by passing a high potential charging bar across the surface of the plate at a uniform velocity. The selenium has good insulation properties and will retain the charge. The sensitized (charged) plate is then placed in a light tight cassette or holder and used in X-ray exposures in the same way as film.

Exposure

During exposure, the X-rays cause the insulating properties of the selenium to break down and the charge leaks through (discharges) to the backing plate. Because the amount of discharge is determined by X-ray intensity, an image of the test object remains on the plate in the form of charged and discharged areas in various degrees. The plate is developed after exposure by spraying it with light colored, sometimes fluorescent, finely divided charged powders, which cling to the charged areas in amounts determined by the degree of charge. The powder coating duplicates the X-ray image, as shown in Figure 8.4.

Figure 8.4: Sample xeroradiograph.

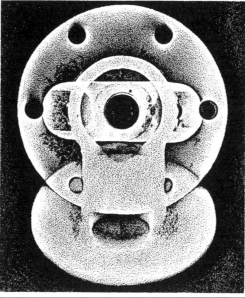

Transfer Process

If a permanent record is desired of a xeroradiograph, the image may be photographed or transferred to a special adhesive white paper. The transfer process uses paper coated with a plastic adhesive. When the paper is pressed on the xeroradiograph, it lifts the powder image from the selenium plate. The image is permanently affixed to the paper by applying sufficient heat to soften the plastic coating and then permitting it to cool.

STEREORADIOGRAPHY AND DOUBLE EXPOSURE

A single radiographic image has length and width, but it does not have perspective. When it is necessary to know the depth of a discontinuity in a thick test object, two radiographic methods are available, stereoradiography and double exposure (parallax).

Stereoradiography

Stereoradiography gives the viewer a three dimensional effect using two radiographs of the test object and a stereoscope. The two radiographs are made with two different positions of the X-ray tube in relation to the test object. The two positions are displaced from each other by a distance equal to the separation of a human's eyes. The stereoscope, through optical means, permits the radiographer to view the two radiographs simultaneously while allowing each eye of the radiographer to see only one of the radiographs. The right eye sees the image of the right shift position of the X-ray tube, and the left eye sees the image of the left shift position. The brain combines and merges the two images into one in which true perspective and spatial relationships are apparent. Stereoradiography is not used in industrial radiography but is of value in discontinuity location or structural visualization, shown in Figure 8.5.

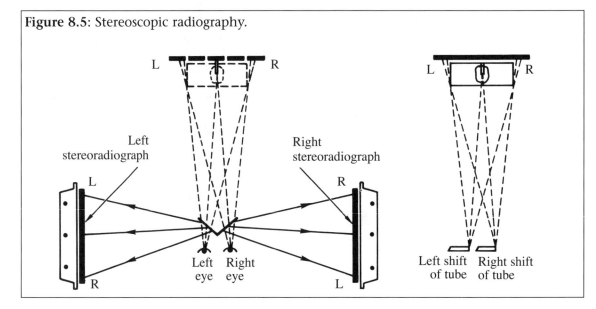

Figure 8.5: Stereoscopic radiography.

Double Exposure (Parallax)

Double exposure (parallax) methods of determining discontinuity depth in a test object are more positive than stereoradiography because they are based on physical measurements of the radiographic image and do not depend on human depth perception.

One such method is illustrated in Figure 8.6. Lead markers M_1 and M_2 are respectively attached to the front and back surfaces of the test object. Two exposures, each one about one half the time required for a normal exposure, are made. The distance between F_1 and F_2 is predetermined, and the tube is located at F_1 for one exposure and at F_2 for the other. The position of the film image of the discontinuity and of M_1 will perceptibly change as a result of the tube shift, while the M_2 image shift will be small if not imperceptible. The distance of the discontinuity from the film plane is determined by the following equation.

Eq. 8.1 $d = \dfrac{bt}{a+b}$

where d is the distance of the discontinuity from the film plane, a is the distance of tube position shift, b is the change in position of the discontinuity image and t is the focus-to-film distance.

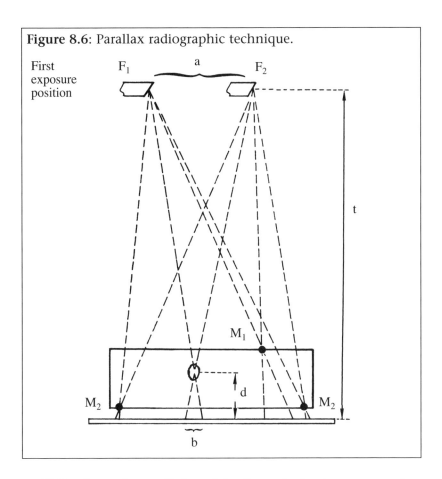

Figure 8.6: Parallax radiographic technique.

If film fog or the small size of the discontinuity does not permit the double exposure technique, two separate radiographs may be made. The two radiographs are aligned by superimposing images of the M_2 markers, the change in position of the discontinuity image is measured, and Eq. 8.1 is applied.

For discontinuity depth determination, when all that is required is knowledge of which test object surface the discontinuity is nearer to, the relationship between the image shift of the M_1 marker and the image shift of the discontinuity provides the answer. If the discontinuity image shift is greater than half the shift of the M_1 image, the discontinuity is closer to the top surface of the test object; if less than half, the discontinuity is closer to the bottom surface.

FLASH RADIOGRAPHY

Flash radiography permits the observation of high speed events in opaque materials. It is used primarily for observation of explosive or rupture processes. Analogous to flash photography, flash radiography freezes the motion of projectiles or high speed machinery by high voltage, high current, extremely short time duration exposures.

The tube and the high voltage circuits of flash radiography equipment differ in design from conventional X-ray equipment. The tube has a cold cathode, and electron emission is initiated by a third electrode located near the cathode. The high voltage circuit contains capacitors that are charged to peak voltage and then discharged in a high voltage pulse. Tube current reaches as high as 2000 amp, but, because of the microsecond duration of the exposure, the tube is not damaged.

IN-MOTION RADIOGRAPHY

In-motion radiography is any radiographic method wherein the source of radiation, the test object or the film is moving during the exposure. Many special in-motion radiographic techniques are in use, each of them designed to serve a specific purpose and application. These techniques use mechanical arrangements to move the X-ray machine, the test object or, in many cases, motion picture cameras loaded with X-ray film.

The one requirement for in-motion radiography is that during exposure the position of the film and the test object relative to each other must remain fixed. This requirement is met by synchronizing the movement of the test object and the film or by fixing the test object and film in position and moving only the source of radiation. The multiple variations of in-motion technique are all based on the requirements mentioned and on the calculations and procedures discussed earlier.

CONCLUSION

Nondestructive testing students completing this classroom training book will have taken an important step, from unaided visual testing to that of a nondestructive testing method that is capable of illuminating discontinuities invisible to the unaided eye.

Additionally, this classroom training book provides background on critical practices for control and implementations of applied radiographic technology.

Students may add technical knowledge in other nondestructive testing technologies to this basic understanding of process qualifications and controls. The results are the uniform and repeatable tests that are a hallmark of an individual certified for radiographic testing.

Glossary

Absorbed dose: Amount of energy imparted to human tissue or a biological system by an ionizing event per unit mass of irradiated material at the place of interest. Absorbed dose is expressed in gray (Gy) or rad.

Absorption: Event where photons in a beam of radiation interact with atoms of a material the photons pass through and are reduced in energy by this interaction.

Accelerator: (1) Device that accelerates charged particles to high energies. Examples are X-ray tubes, linear accelerators and betatrons. (2) Linear accelerator.

Acceptable quality level (AQL): Maximum percentage of defective units of the total units tested in an acceptable lot.

Acceptance criteria: Standard against which test results are to be compared for purposes of establishing the functional acceptability of a test object or system being tested.

Acceptance level: Average or standard criteria above or below which test objects are acceptable, in contrast to rejection level.

Acceptance standard: Specimen similar to the test object containing natural or artificial discontinuities that are well defined and similar in size or extent to the maximum acceptable in the product. See **Standard**.

Accommodation: Of the eye, adjustment of the lens' focusing power by changing the thickness and curvature of the lens by the action of tiny muscles attached to the lens. Accommodation facilitates the viewing of objects near and far.

Accuracy: Degree of conformity of a particular measurement to a standard or true value.

Activity: Degree of radioactivity of a particular isotope. Activity is expressed as the number of atoms disintegrating per unit of time. Measured in becquerel (curie).

Agency: Organization selected by an authority to perform nondestructive testing, as required by a specification or purchase order.

Algorithm: Prescribed set of well defined rules or processes for the solution of a mathematical problem in a finite number of steps.

Alpha particle: Positively charged ion emitted by certain radioactive materials. It is made up of two neutrons and two protons; hence it is identical with the nucleus of a helium atom.

Alternating current: Electrical current that reverses its direction of flow at regular intervals.

Alternating magnetic field: Varying magnetic field produced around a conductor by an alternating current flowing in the conductor.

Ampere (A): Unit of electric current.

Analog-to-digital converter: Circuit whose input is information in analog form and whose output is the same information in digital form.

Angstrom (Å): Unit of distance once used to express wavelengths of electromagnetic radiation. The SI unit nanometer (nm) is now preferred; 1 nm = 10 Å.

Anode: (1) In radiography, the positive electrode of a cathode ray tube that generates ionizing radiation. (2) Positively charged terminal, which may corrode electrochemically during production of an electric current. Compare **Cathode**.

ASNT: The American Society for Nondestructive Testing, Inc.

ASNT *Recommended Practice No. SNT-TC-1A*: See *Recommended Practice No. SNT-TC-1A*.

Attenuation: The decrease in radiation intensity caused by distance and by passage through material.

Automated system: Acting mechanism that performs required tasks at a determined time and in a fixed sequence in response to certain conditions. All called *robotic system*.

Backscatter: (1) Interaction of radiation with matter such that the direction of travel after scattering is over 90° and often close to 180° to the original direction of travel. (2) In transmission radiography, interaction of radiation with matter behind the image plane such that scattered radiation returns to the image plane, often adding fog and noise that interfere with production of an image of the test object. (3) Of scatter imaging, interaction of incident radiation with a test object that scatters the radiation through large angles frequently greater than 90° to the original direction of travel. Such radiation is used to form an image or to measure a parameter of the test object, usually through digital techniques.

Backscatter imaging: In radiographic testing, a family of techniques that use backscatter for image generation.

Barium clay: Molding clay containing barium, used to eliminate or reduce the amount of scattered or secondary radiation reaching the film.

Beam: Defined stream of radiation in which all elements are traveling in nearly parallel paths.

Beam quality: Penetrating energy of a radiation beam.

Beam spread: Divergence from a beam of radiation in which all elements are traveling in parallel paths.

Becquerel (Bq): SI unit for measurement of radioactivity, equivalent to one disintegration per second. Replaces curie. One curie equals 37 GBq.

Beta particle: Electron or positron emitted from a nucleus during decay.

Beta ray: Radiation stream consisting of beta particles.

Betatron: Circular electron accelerator that is a source of either high energy electrons or X-rays. The electrons are injected by periodic bursts into a region of an alternating magnetic field. Sometimes the electrons are used directly as the radiation.

Billet: Solid semifinished round or square product that has been hot worked for forging, rolling or extrusion.

Bleed: Refers to molten metal oozing out of a casting. Stripped or removed from the mold before complete solidification.

Brazing: Joining of metals and alloys by fusion of nonferrous alloys that have melting points above 840 °F (449 °C), but below melting points of materials being joined.

Brehmsstrahlung: Electromagnetic radiation produced when electrons' path and kinetic energy brings them close to the positive fields of atomic nuclei — as when electrons strike a target provided for this purpose. The electrons slow down, giving up kinetic energy as X-radiation.

Burning: Extreme overheating. Makes metal grains excessively large and causes the more fusible constituents of steel to melt and run into the grain boundaries or it may leave voids between the grains. Steel may be oxidized to the extent that it is no longer useful and cannot be corrected by heat treating but it can be remelted.

Burst: In metal, external or internal rupture caused by improper forming.

Butt weld or **butt joint**: Weld joining two metal pieces in the same plane.

Camera: Device that contains a sealed radiation source, where the source or shielding can be moved so that the source becomes unshielded (to make a radiographic exposure) or shielded (for safe storage).

Casing: Many strings of pipe that are used to line the hole during and after drilling of a gas or oil well.

Cassette, film: Lightproof container that is used for holding radiographic film in position during the radiographic exposure. The cassette may be rigid or flexible and may contain intensifying screens, filter screens, both or neither.

Casting: Object of shape obtained by solidification of a substance in a mold.

Casting shrinkage: Total shrinkage includes the sum of three types: (1) liquid shrinkage (the reduction in volume of liquid metal as it cools through the liquidus to the solidus); (2) solidification shrinkage (the change in volume of metal from the beginning to ending of solidification); and (3) solid shrinkage (the reduction in volume of metal from the solidus to room temperature).

Cathode: (1) Negatively charged terminal in an arrangement that produces current by chemical reactions. Compare **Anode**. (2) In radiography, the negative electrode of an X-ray tube, the electrode from which electrons are emitted.

Cathode ray: Stream of electrons emitted by a heated filament and projected in a more or less confined beam under the influence of a magnetic or electric field.

Certification: Process of providing written testimony that an individual (or test technique, process or equipment) is qualified. See also **Certified**.

Certified: Having written testimony of qualification. See also **Certification**.

Cesium-137 (Cs-137): Radioactive isotope of element cesium, having a half life of about 30 years and photon energy of about 660 keV.

Clean: Free from interfering solid or liquid contamination on the test surface and within voids or discontinuities.

Cobalt-60 (Co-60): Radioactive isotope of element cobalt, having half life of 5.3 years and photon energies of 1.17 and 1.33 MeV.

Code: A written standard enacted or enforced as a law.

Compton scatter: Reduction of the energy of an incident photon by its interaction with an electron. Part of the photon energy is transferred to the electron, giving it kinetic energy, and the remaining photon is redirected with reduced energy.

Contrast: (1) In film radiography, the measure of differences in the film blackening or density resulting from various radiation intensities transmitted by the object and recorded as density differences in the image. Thus, difference in film blackening from one area to another. (2) The difference in visibility between an indication and the surrounding area.

Contrast, subject: Ratio of radiation intensities transmitted by selected portions of the object being radiographed.

Control: See **Process control** and **Quality control**.

Corrosion: Deterioration of a metal by chemical or electrochemical reaction with its environment. Removal of material by chemical attack, such as the rusting of automobile components.

Crack: (1) Break, fissure or rupture, usually V shaped (the cross section view that otherwise appears jagged) and relatively narrow and deep. A discontinuity that has a relatively large cross section in one direction and a small or negligible cross section when viewed in a direction perpendicular to the first. (2) Propagating discontinuities caused by stresses such as heat treating or grinding. Difficult to detect unaided because of fineness of line and pattern (may have a radial or latticed appearance).

Curie (Ci): Unit of measurement of the quantity of radioactivity. Replaced by becquerel in SI, where 1 Ci = 3.7×10^{10} Bq, or 1 Ci = 37 GBq.

Decay curve: Graph showing radioactive strength in becquerel (curie) as a function of time for an isotope. Decay curves are used in determining exposure times in radiographic testing.

Defect: Discontinuity that exceeds the acceptance criteria or is detrimental to the service of the test object. See also **Discontinuity**.

Definition: Description of linear demarcation sensitivity, or the detail sharpness of object outline in a radiographic image. It is a function of screen type, exposure geometry, radiation energy and characteristics of film or sensor.

Delamination: Laminar discontinuity, generally an area of unbonded materials.

Depth of field: Range of distance over which an imaging system gives satisfactory definition when its lens is in the best focus for a specific distance.

Depth of focus: Distance a sensor may be moved from a lens system and still produce a sharp image.

Depth of fusion: Depth to which base metal has melted during welding.

Detail: In radiography, the degree of sharpness of outline of an image, or the clear definition of an object or discontinuity in the object. See also **Definition**.

Developer: In radiography, a chemical solution that reduces exposed silver halide crystals to black metallic silver.

Diffraction: A special case of scatter, where coherently scattered photons undergo interference or reinforcement, resulting in patterns indicative of the scattering medium. See also **X-ray diffraction**.

Discontinuity: Unintentional interruption in the physical structure or configuration of a test object. After nondestructive testing, discontinuities interpreted as detrimental in the host object may be called *defects*.

Discontinuity, inherent: Material anomaly originating from solidification of molten metal. Pipe and nonmetallic inclusions are the most common and can lead to other types of discontinuities in fabrication.

Dose: See **Absorbed dose**.

Dose rate: Radiation dose delivered during a specified unit of time and measured, for instance, in sievert per minute (or in rem per hour). See also **Absorbed dose**.

Dosimeter: Device that measures radiation dose, such as an ionization chamber.

Effective focal spot: Size and geometry of focal spot after target interaction. Viewed from along the primary beam central axis at the target the effective focal spot would appear nearly square and smaller than the actual focal spot area covered by the electron stream.

Evaluation: Process of determining the magnitude and significance of a discontinuity after the indication has been interpreted as relevant. Evaluation determines if the test object should be rejected, repaired or accepted. See **Indication** and **Interpretation**.

Exposure factor: In X-radiography, the quantity that combines source strength (milliampere), time (usually minute) and distance. It is the product of milliamperage and time divided by distance squared and determines the degree of film density.

Field: In video technology, one of two video picture components that together make a frame. Each picture is divided into two parts called *fields* because a frame at the rate of thirty frames per second in a standard video output would otherwise produce a flicker discernible to the eye. Each field contains one half of the total picture elements. Two fields are required to produce one complete picture or frame so the field frequency is sixty fields per second and the frame frequency is thirty frames per second.

Field of view: Range or area that can be seen through an imaging system, lens or aperture.

Film badge: Package of photographic film worn as a badge by radiographic personnel (and by workers in the nuclear industry) to measure exposure to ionizing radiation. Absorbed dose can be calculated by degree of film darkening caused by irradiation.

Film holder: See **Cassette, film**.

Film speed: Relative exposure required to attain a specified film density.

Filter: (1) Network that passes electromagnetic wave energy over a described range of frequencies and attenuates energy at all other frequencies. (2) Processing device or function that excludes a selected kind of signal or part of a signal. (3) In radiography, the thickness of absorbing material placed in a primary radiation beam to selectively remove longer wavelength radiation, thereby adjusting the quality of the radiographic image.

Fixing: Procedure used in film processing that removes undeveloped silver salts in the emulsion from the surface of the film, leaving only the developed black silver of the image on the film.

Flakes: Short discontinuous internal fissures in ferrous metals attributed to stresses produced by localized transformation and/or decreased solubility of hydrogen during cooling usually after hot working. On a fractured surface, flakes appear as bright silvery areas; on an etched surface they appear as short, discontinuous cracks. Also called *shatter cracks* and *snowflakes*.

Focal spot: Area on target that receives bombardment of electrons. See also **Effective focal spot**.

Focus: Position of a viewed object and a lens system relative to one another to offer a distinct image of the object as seen through the lens system. See **Accommodation** and **Depth of field**.

Focus, principal plane of: Single plane actually in focus in a photographic scene.

Fog: Increase of film density caused by sources other than from the intended primary beam exposure. Heat, humidity, pressure and scatter radiation can all cause fogging of the film.

Fracture: Break, rupture or crack large enough to cause a full or partial partition of a casting.

Frame: Complete raster scan projected on a video screen. There are thirty frames per second in a standard video output. A frame may be comprised of two fields, each displaying part of the total frame. See also **Field**.

Gamma rays: High energy, short wavelength electromagnetic radiation emitted by the nucleus of a radioactive isotope. Energies of gamma rays are usually between 0.01 and 10 MeV. X-rays also occur in this energy range but are of nonnuclear origin.

Geometric unsharpness: See **Unsharpness, geometric**.

Gray (Gy): SI unit for measurement of the dose of radiation absorbed per unit mass at a specified location. Replaces the rad where rad denotes radiation absorbed dose, not radian.
1 Gy = 1 J·kg^{-1} = 100 rad.

Gray level: Integer number representing the luminance or darkness of a pixel or, as a composite value, of an image comprised of pixels.

Image: Visual representation of a test object or scene.

Image enhancement: Any of a variety of image processing steps, used singly or in combination to improve the detectability of objects in an image.

Image processing: Actions applied singly or in combination to an image, in particular the measurement and alteration of image features by computer. Also called *picture processing*.

Image quality indicator: Strip of material the same composition as that of the material being tested, representing a percentage of object thickness and provided with a combination of steps, holes or slots or alternatively made as a series of wires. When placed in the path of the radiation, its image provides a check on the radiographic technique used.

In-motion radiography: Technique in which either the object being radiographed or the source of radiation is in motion during the exposure.

Index of refraction: Ratio of velocity of light in a vacuum to velocity of light in a material.

Indication: Nondestructive testing response that requires interpretation to determine its relevance.

Indication, discontinuity: Visible evidence of a material discontinuity. Subsequent interpretation is required to determine the significance of an indication.

Indication, false: (1) Indication produced by something other than a discontinuity or test object configuration. (2) Indication caused by misapplied or improper technique.

Indication, nonrelevant: Indication caused by a condition that does not affect the usability of the object (a change of section, for instance).

Indication, relevant: Indication from a discontinuity (as opposed to a nonrelevant indication) requiring evaluation by a qualified technician, typically with reference to an acceptance standard, by virtue of the discontinuity's size or location.

Inherent discontinuities: Discontinuities that are produced in the material at the time it is formed (for example, during solidification from the molten state).

Interpretation: Determination of the significance of nondestructive testing indications from the standpoint of their relevance or nonrelevance.

Inverse square law: From a point source of radiation, the intensity of energy decreases as the inverse square of distance from the source increases and vice versa.

Ionizing radiation: Form of radiation that can displace orbital electrons from atoms. Types include X-rays, gamma rays and particles such as neutrons, electrons and alpha particles.

IQI: See **Image quality indicator**.

Iridium-192 (Ir-192): Radioactive isotope of the element iridium, having a half life of 73 to 75 days and primary photon energies of 0.31, 0.47 and 0.66 MeV.

Irradiance: Power of electromagnetic radiant energy incident on the surface of a given unit area. Compare **Radiance**.

Level, acceptance: In contrast to rejection level, test level above or below which, depending on the test parameter, test objects are acceptable.

Level, rejection: Value established for indication or test signal above or below which, depending on the test parameter, test objects are rejectable or otherwise distinguished from the remaining objects. See **Level, acceptance**.

Linear accelerator: High frequency electron generator.

Material noise: Random signals caused by the material structure of the test object. A component of background noise.

Mechanical properties: Properties of a material that reveal its elastic and inelastic behavior where force is applied, thereby indicating its suitability for mechanical applications (for example, modulus of elasticity, tensile strength, elongation, hardness and fatigue limit).

Milliroentgen: A radiation dose measurement replaced by sievert. 100 000 mR = 1 Sv.

Neutron: Uncharged elementary particle with mass nearly equal to that of the proton.

Neutron radiography: Radiographic testing using a neutron beam.

Neutron radioscopy: Radioscopy using a neutron beam.

Noise: Any undesired signals that tend to interfere with normal detection or processing of a desired signal.

Nondestructive testing (NDT): Determination of the physical condition of an object without affecting that object's ability to fulfill its intended function. Nondestructive testing techniques typically use a probing energy to determine material properties or to indicate the presence of material discontinuities (surface, internal or concealed).

Nonrelevant indication: See **Indication, nonrelevant**.

One hundred percent testing: Testing of all parts of an entire production lot in a prescribed manner. Compare **Sampling, partial**.

Orientation: Angular relationship of a surface, plane, discontinuity or axis to a reference plane or surface.

 : See **Image quality indicator**.

Peripheral vision: Seeing of objects displaced from the primary line of sight and outside the central visual field.

Phase shift: Change in the phase relationship between two alternating quantities of the same frequency.

Photoelectric effect: Emission of free electrons from a surface bombarded by sufficiently energetic photons. Such emissions may be used in an illuminance meter and may be calibrated in lux.

Photoemission: Method by which an image orthicon television camera tube produces an electrical image, in which a photosensitive surface emits electrons when light reflected from a viewed object is focused on that surface.

Photometry: Science and practice of the measurement of light or photon-emitting electromagnetic radiation.

Photon: Quantum of electromagnetic radiation.

Photoreceptor: Photon sensor. Examples include film and electronic detector elements.

Physical properties: Nonmechanical properties such as density, electrical conductivity, heat conductivity and thermal expansion.

Picture element: See **Pixel**.

Picture processing: See **Image processing**.

Pixel: One element of a digital image. Each pixel represents a finite area in the scene being imaged.

Plane of focus: See **Focus, principal plane of**.

Primary radiation: Radiation emitting directly from the target of an X-ray tube or from a radioactive source.

Principal plane of focus: See **Focus, principal plane of**.

Process: Repeatable sequence of actions to bring about a desired result.

Process control: Application of quality control principles to the management of a repeated process.

Process testing: Initial product testing to establish correct manufacturing procedures and then by periodic tests to ensure that the process continues to operate correctly.

Qualification: Process of demonstrating that an individual (or test technique, process or instrument) has the required amount and the required type of training, experience, knowledge and abilities. See also **Qualified**.

Qualified: Having demonstrated the required amount and the required type of training, experience, knowledge and abilities. See also **Qualification**.

Quality: Ability of a process or product to meet specifications or expectations of its users in terms of efficiency, appearance, longevity and ergonomics.

Quality assurance: Administrative actions that specify, enforce and verify a quality control program.

Quality control: Physical and administrative actions required to ensure compliance with the quality assurance program. May include nondestructive testing in the manufacturing cycle.

Rad: Radiation absorbed dose. Unit of absorbed dose of ionizing radiation. One rad is equal to the absorption of 100 erg (10^{-5} J) of radiation energy per gram of matter associated with human tissue or a biological system. Replaced by the gray (Gy).

Radiance: Radiant flux per unit solid angle and per unit projected area of the source. Measured in watts per square meter steradian. Compare **Irradiance**.

Radiant energy: Energy emitting as electromagnetic waves. Also known as *radiation*.

Radiant flux: Radiant energy's rate of flow, measured in watts.

Radiant intensity: Electromagnetic energy emitted per unit time per unit solid angle.

Radiant power: Total radiant energy emitted per unit time.

Radiation safety officer: Individual supervising a program to provide radiation protection. The representative appointed by the licensee for liaison with the applicable regulatory agency.

Radiographer: Person who performs, supervises and is responsible for industrial radiographic testing operations.

Radiographic interpretation: Determination of the cause and significance of indications on a radiograph.

Radiographic screens: Fluorescent sheets or lead used to intensify the effect of radiation on films. The screens can be made of a fluorescent metal. Metallic screens help absorb secondary and scattered radiation, which helps to improve image quality.

Radiographic testing (RT): Penetrating radiant energy in the form of X-rays, gamma rays or neutrons for nondestructive testing of objects to provide images of the objects' interiors. Also called *radiography*.

Radiography: Radiographic testing.

Radiology: (1) That branch of medicine which uses ionizing radiation for diagnosis and therapy. (2) Science of electromagnetic radiation, particularly ionizing radiation.

Radiometer: Instrument for measuring radiant power of specified frequencies. Different radiometers exist for different frequencies.

Radiometric photometer: Radiometer for measuring radiant power over a variety of wavelengths.

Radioscopy: Radiographic testing technique in which gamma rays, X-rays or neutrons are used to produce an image on a video or screen display as opposed to a latent image on a film. The test object or interrogating optics may move in real time to present a moving radiographic image.

Recommended practice: Set of guidelines or recommendations.

Recommended Practice No. SNT-TC-1A: Set of guidelines for employers to establish and conduct a nondestructive testing personnel qualification and certification program. *SNT-TC-1A* was first issued in 1968 by the Society for Nondestructive Testing (SNT, now ASNT) and has been revised every few years since.

Rejection level: See **Level, rejection**.

Relevant indication: See **Indication, relevant**.

Rem: Roentgen equivalent man. Unit of absorbed radiation dose in biological matter. It is equal to the absorbed dose in rads multiplied by the quality factor of the radiation.

Repeatability: Ability to reproduce a detectable indication in separate processings and tests from a constant source.

Residual elements: Elements present in an alloy in small quantities, but not added intentionally.

Resolution: Aspect of image quality pertaining to a system's ability to reproduce objects, often measured by resolving a pair of adjacent objects or parallel lines.

Resolution, discontinuity: Property of a test system that enables the separation of indications caused by discontinuities located in close proximity to each other in a test object.

Resolution test: Procedure wherein a line or a series of lines or line pairs are detected to verify or evaluate a system's sensitivity.

Resolution threshold: Minimum distance between a pair of points or parallel lines when they can be distinguished as two, not one, expressed in minutes of arc. Vision acuity in such a case is the reciprocal of one half of the period expressed in minutes.

Resolving power: Ability of detection systems to separate two points in time or distance. Resolving power depends on the angle of vision and the distance of the sensor from the test surface. Resolving power in vision systems is often measured using parallel lines. Compare **Resolution**.

Roentgen (R): Unit for measurement of radiation intensity; amount of radiation that will generate one electrostatic unit in 1 cm^{-3} of air at standard atmospheric conditions. The roentgen (R) has been replaced by an SI compound unit, coulomb per kilogram ($C \cdot kg^{-1}$).

Sampling, partial: Testing of less than 100% of a production lot. See also **One hundred percent testing**.

Sampling, random partial: Partial sampling that is fully random.

Sampling, specified partial: Partial sampling in which a particular frequency or sequence of sample selection is prescribed. An example of specified partial sampling is the testing of every fifth unit.

Scattering: Random reflection and refraction of radiation caused by interaction with material it strikes or penetrates.

Sensitivity: Measure of a sensor's ability to detect small signals. Limited by the signal-to-noise ratio.

Sensor, X-ray: In radiographic testing, device or material that changes with and provides evidence of contact with ionizing radiation. Examples include X-ray film, X-ray sensitive phosphors and electronic devices such as linear detector arrays.

Shielding: Material or object used to reduce intensity of or exposure to penetrating radiation.

SI: International System of units of measurement. An international system of measurement based on seven units: meter (m), kilogram (kg), second (s), kelvin (K), ampere (A), candela (cd) and mole (mol).

Sievert (Sv): SI unit for measurement of exposure to ionizing radiation, replacing rem. 1 Sv = 1 J·kg^{-1} = 100 rem.

Signal: Response containing relevant information.

Signal processing: Acquisition, storage, analysis, alteration and output of digital data through a computer.

Signal-to-noise ratio: Ratio of signal values (responses that contain relevant information) to baseline noise values (responses that contain nonrelevant information). See **Noise**.

Source: Machine or material from which ionizing radiation emanates.

Specification: Set of instructions or standards to govern the results or performance of a specific set of tasks or products.

Spectrum: (1) Amplitude distribution of frequencies in a signal. (2) Representation of radiant energy in adjacent bands of hues in sequence according to the energy's wavelengths or frequencies. A rainbow is a well known example of the visible light spectrum.

Spectrum response: Amplification (gain) of a receiver over a range of frequencies.

Spot check tests: Testing a number of objects from a lot to determine the lot's quality, the sample size being chosen arbitrarily, such as 5 or 10%. This does not provide accurate assurance of the lot's quality.

Spot examination: Local examination of welds or castings.

Standard: (1) Physical object with known material characteristics used as a basis for comparison, specification or calibration. (2) Concept established by authority, custom or agreement to serve as a model or rule in the measurement of quantity or the establishment of a practice or procedure. (3) Document to control and govern practices in an industry or application, applied on a national or international basis and usually produced by consensus. See also **Acceptance standard**.

Step wedge: See **Stepped wedge**.

Stepped wedge: Reference object, with steps of various thicknesses in the range of the test objects' thicknesses, for the radiographic testing of objects having thickness variations or complex geometries. The stepped wedge must be made of material radiographically similar to that of the radiographic test object and may include image quality indicator features (such as calibrated holes) in any or all steps.

Stereo imaging: Imaging technique involving the capture and display of two images of the same object from different angles. Binocular viewing simultaneously of the two images simulates a three dimensional viewing.

Stereoradiography: Radiographic testing using stereo imaging.

Survey meter: Portable instrument that measures rate of exposure dose or ionizing radiation intensity.

Threshold level: Setting of an instrument that causes it to register only those changes in response greater or less than a specified magnitude.

Tolerance: Permissible deviation or variation from exact dimensions or standards.

Unsharpness, geometric: Fuzziness or lack of definition in a radiographic image resulting from the source size, object-to-film distance and the source-to-object distance.

Video: Pertaining to the transmission and display of images in an electronic format that can be displayed on a screen.

Video presentation: Electronic screen presentation in which radiofrequency signals have been rectified and usually filtered.

X-ray: Penetrating electromagnetic radiation emitted when the inner orbital electrons of an atom are excited and release energy. Radiation is nonisotopic in origin and is generated by bombarding a metallic target with high speed charged particles, usually electrons.

X-ray diffraction: Radiographic testing technique used for material characterization, based on change in scattering of X-radiation as a result of interaction with test material. See also **Diffraction**.

X-ray fluorescence: Radiographic testing technique used for material characterization, based on wavelengths of fluorescence from material irradiated by X-rays.

Bibliography and Figure Sources

BIBLIOGRAPHY

1. *Nondestructive Testing Handbook*, third edition: Vol. 4, *Radiographic Testing*. Columbus, Ohio: American Society for Nondestructive Testing (2002).

2. *Nondestructive Testing Handbook*, second edition: Vol. 3, *Radiography and Radiation Testing*. Columbus, Ohio: American Society for Nondestructive Testing (1985).

3. *Nondestructive Testing Classroom Training Handbook*, second edition: *Radiographic Testing*. Fort Worth, Texas: Convair Division of General Dynamics Corporation (1983).

4. *Working Safely in Radiography*. Columbus, Ohio: American Society for Nondestructive Testing (2004).

5. *Nondestructive Testing*. Materials Park, Ohio: ASM International (1995).

FIGURE SOURCES

The following credits indicate the sources of illustrations in this book. All figures reprinted with permission.

Chapter 2
Figures 2.1 to 2.5, 2.8 to 2.18: Reprinted from *Nondestructive Testing Classroom Training Handbook*, second edition: *Radiographic Testing*.

Figures 2.6 and 2.7: Reprinted from *Nondestructive Testing Handbook*, third edition: Vol. 4, *Radiographic Testing*.

Chapter 3
Figures 3.1 and 3.2: Reprinted from *Nondestructive Testing Classroom Training Handbook*, second edition: *Radiographic Testing*.

Figure 3.3: Reprinted from *Nondestructive Testing Handbook*, second edition: Vol. 3, *Radiographic Testing*.

Figures 3.4 to 3.6: Reprinted from *Nondestructive Testing Handbook*, third edition: Vol. 4, *Radiographic Testing*.

Chapter 4
Figures 4.1 to 4.6: Reprinted from *Nondestructive Testing Classroom Training Handbook*, second edition: *Radiographic Testing*.

Chapter 5
Figures 5.1, 5.4 and 5.5: Reprinted from *Nondestructive Testing Classroom Training Handbook*, second edition: *Radiographic Testing*.

Figures 5.2 and 5.3: Courtesy of the United States Nuclear Regulatory Commission.

Figure 5.6: Reprinted from *Nondestructive Testing Handbook*, third edition: Vol. 4, *Radiographic Testing*.

Chapter 6
Figures 6.1 to 6.4, 6.6 to 6.40: Reprinted from *Nondestructive Testing Classroom Training Handbook*, second edition: *Radiographic Testing*.

Figure 6.5: Courtesy of Jerry Fulin, El Paso Corporation.

Chapter 7
Figures 7.1 to 7.9: Reprinted from *Nondestructive Testing Handbook*, third edition: Vol. 4, *Radiographic Testing*.

Chapter 8
Figures 8.1 to 8.6: Reprinted from *Nondestructive Testing Classroom Training Handbook*, second edition: *Radiographic Testing*.

Index

A

absorption, of radiation, 11, 15
  as exposure variable, 102
accelerators, 32-35
acceptance criteria, 116
accessory equipment, 82-98
ACCP-ASNT Central Certification Programs, 8-9
acetic acid, for stop bath, 51
activity, of radioisotopes, 26, 27
aerospace industry, radiography applications, 3
afterglow, 144
agreement states, 57
airborne radioactivity area (sign), 72
allowable working time, 61
alpha particles, 26
  quality factor value, 58*table*
aluminum
  amorphous silicon detector imaging of weld, 147, 149
  example exposure calculations, 109-111
  practical thickness limits, 98*table*
  for radiation shielding, 64*table*
  radiographic equivalent factor, 97, 98*table*
  X-ray exposure charts, 103
ambient dose equivalent, 60
American Society for Nondestructive Testing (ASNT)
  ACCP-ASNT Central Certification Programs, 8-9
  Recommended Practice No. SNT-TC-1A, 4-5, 7
amorphous selenium detectors
  afterglow, 144
  principles of, 137, 138
  technology, 149
amorphous silicon detector array, 146, 147
amorphous silicon detectors
  principles of, 137, 138
  technology, 146-149
angular measuring devices, 89
annular X-ray beam configuration, 33
anode, 16, 17, 31, 34
  heat dissipation, 35

ANSI/ASNT CP-189-1991, 5
ANSI/ASNT CP-189-2001, 5
area alarm systems, 77
area shielding equipment, 91, 101
artificial gamma ray sources, 26
ASNT, See *American Society of Nondestructive Testing*
ASNT CP-189, 5
atomic mass unit, 21*table*
atomic particles, 21*table*
atomic reactors, gamma ray sources produced in, 26
atoms, 21-22
automatic film processing, 53-54
automatic processing darkroom, 55
automotive manufacture, X-radiography linear detector
      array application, 151

B

backscatter radiation, 25
  controlling, 84, 101
barium clay, as masking material, 86
barium fluorobromide, europium activated, 139
barrels, X-radiography linear detector array application,
      151, 152
base fog, 44
becquerel (Bq), 59
beta particles, 26
  quality factor value, 58*table*
bonded honeycomb radiography, 127
braking radiation, 16
brass
  cobalt-60 for gamma radiography of, 38
  radiographic equivalent factor, 98*table*
brass filters, 83
brazed honeycomb radiography, 127-130
brick, for radiation shielding, 67
bucket blade radioscopic imaging, 151
buildup, 24
butt welds, radiography, 117

C

calcium tungstate, fluorescent screens made from, 84, 100

camera (isotope camera), 40

capacitors, radiography, 132, 133

cargo transport, X-radiography linear detector array
    application, 151

cassettes, 89

castings, radiography, 3

cathode, 16, 17, 30-31, 34

caution signs, 72

ceilings, of darkrooms, 55

certification, for radiographic testing, 5, 7-9

cesium-137, gamma ray source, 26, 39-40
    dose rate at 1 ft emissivity, 62*table*
    dose rate per curie versus distance, 63*table*
    gamma ray energy, 28
    half value layers, 64*table*
    radiographic equivalent factors using, 98*table*
    radioisotope characteristics, 40*table*
    voltage versus thickness range for steel, 108

cesium iodide phosphors, 144, 147, 148

chalk, for marking, 90

characteristic curves, of radiographic film, 44-47, 97-98

characteristic X-ray, 15-16, 18-19

charge coupled devices (CCDs), 135, 137-138
    applications, 144-145
    development of, 136
    technology, 150-151

Civil Aeronautics Board (CAB), enforcement of
    radioactive material regulations, 57

cleanliness, importance in film developing, 49, 56

closed sphere radiography, 120, 121

closed tank radiography, 122

cobalt-60, gamma ray source, 26, 38-39
    dated decay curve, 97
    decay rate, 38*table*
    dose rate at 1 ft emissivity, 62*table*
    dose rate per curie versus distance, 63*table*
    exposure charts, 95, 96
    fluorescent screens used with, 85
    gamma ray energy, 28
    half value layers, 64*table*
    radiographic equivalent factors using, 98*table*
    radioisotope characteristics, 40*table*
    voltage versus thickness range for steel, 108

collimators, 83, 101
    in linear accelerators, 33

color visual test, 7

committed dose equivalent, 60

compton effect, 23, 28

concrete, for radiation shielding, 64, 64*table*, 66, 67

cones, 83, 101

continuous X-ray radiation, 16, 18-19

contrast, 42, 43-47, 102

copper
    as anode material, 31
    cobalt-60 for gamma radiography of, 38
    radiographic equivalent factor, 98*table*

copper filters, 83

corner joints, radiography of welded, 118

coulumb per kilogram, 59-60

cracks
    detector resolution for successful, 141
    radiograph, 42
    radiography used for locating, 3

critical criteria, 116

curie (Ci), 26, 59

D

danger radioactive material-do not handle (tag), 72

darker film images, 11

darkroom equipment, 55

darkroom facilities, 54-55

dated decay curves, 27, 97

decay curve, 27

deep dose equivalent, 60

definition, 42

densitometer, 91

Department of Transportation (DOT), enforcement of
    radioactive material regulations, 57

detection and measurement instruments, for radiation, 73-77

detectors, for digital radiographic imaging, See *digital
    radioscopic detectors*

developer, 49, 50-51

developer tanks, 55

developing, of radiographic film, 50-51

diaphragms, 83, 101

differential absorption, of radiation, 11

digital radiographic imaging, 135-136, 135-152

detection efficiency, 140

selection of systems, 144-146

spatial resolution, 141-146

digital radioscopic detectors, 136
    principles, 137-140
    technology, 146-152

diodes, radiography, 131, 132, 133

direct conversion, 149

directional dose equivalent, 60

J

joints, radiography of welded corner, 118

K

kilo-electron volts (keV), 16
kilovoltage, exposure variable, 101

L

labels, 72
lack of fusion, radiograph, 42
lag, 144
lasers, in linear accelerators, 33
lateral conical X-ray beam configuration, 33
lead
    as masking material, 86
    for diaphragms, collimators, and cones, 83
    for enclosed exposure areas, 65
    for radiation shielding, 64, 64*table*, 66-67, 91
    for X-ray tube shielding, 65
    radiographic equivalent factor, 98*table*
lead filters, 83
lead screens, 84, 85-86, 98*table*
    for controlling scatter, 100-101
lead sheet shields, 91, 101
lens, as optical coupling for CCDs, 151
lens dose equivalent, 60
Level I personnel, 5-6
    examination, 8
Level II personnel, 5, 6
    examination, 8
    examination for ACCP-ASNT Central Certification
      Program, 9
Level III personnel, 5, 6
    examination, 8
    examination for ACCP-ASNT Central Certification
      Program, 9
light, protection against outside in darkrooms, 54
lighter film images, 11
linear accelerators, 32-35
linear array detectors
    applications, 145
    development of, 136
    principles of, 137, 139
    technology, 151-152
linear measuring devices, 89
liquid developer, 51
liquid fixer, 52
location markers, 90
longitudinal cracks, radiograph, 42

low speed film, 46, 47
luminance amplification factor, 154

M

magnesium, radiographic equivalent factor, 98*table*
masking material, 86, 101
matter interaction
    gamma rays, 28
    X-rays, 21-25
maximum permissible dose, 60-61, 71*table*
medical digital X-ray imaging, 135-136
medium speed film, 46
metallic shot, as masking material, 86
metal oxide semiconductor (MOS) capacitors, 150
milliamperage, exposure variable, 101
million-electron volts (MeV), 16
milliroentgen (mR), 58
modulation transfer function, 142
movement, exposure variable, 98-99
multiple combination application, 123
munitions, X-radiography linear detector array
    application, 151

N

neutrons, 21, 21*table*
    quality factor value of thermal and fast, 58*table*
nickel, as cathode material, 30
nickel alloy, bucket blade radioscopic imaging, 151
noncritical criteria, 116
nondestructive testing, 6
NRC Form-4: Cumulative Occupational Dose History,
    68, 69, 71
*NRC Form-5: Occupational Dose Record for a
    Monitoring Period*, 68, 70, 71
nuclear reactors, gamma ray sources produced in, 26
Nuclear Regulatory Commission (NRC), 38, 57, 68-73
nuclear waste containment, X-radiography linear detector
    array application, 151
number belts, 90

O

object-to-film distance
    exposure variable, 99
    and sharpness, 13, 14
occupational annual dose limits, 60-61
occupational radiation exposure limits, 68, 71
offset correction, radioscopic detection systems, 143
online training, 8

---

radiographic testing, 3-4. See also *digital radiographic imaging*; *gamma radiography*; *safety*; *X-radiography*
  advantages, 3
  applications, 115-133
  basic process, 11
  brazed honeycombs, 127-130
  certification, 5, 7-9
  closed spheres, 120, 121
  closed tanks, 122
  digital imaging, 135-152
  disadvantages, 3-4
  discontinuity location techniques, 126-127
  double exposure (parallax) radiography, 158-159
  equipment selection, 81-82
  flash radiography, 159-160
  fluoroscopy applications, 133, 153-154
  hemispherical sections, 124
  image amplifier, 154-155
  image enlargement, 12
  image sharpness, 13-14
  in-motion, 160
  introduction, 3-9
  large pipe welds, 125
  Level I personnel, 5-6, 8
  Level II personnel, 5-6, 8, 9
  Level III personnel, 5-6, 8, 9
  limitations, 3-4
  multiple combination application, 123
  panoramic, 124-125
  parallax radiography, 158-159
  principles, 11-28
  qualification, 4-6, 7
  semiconductors, 131-133
  special techniques, 153-160
  stereoradiography, 157-158
  television, 155-156
  test objective, 4
  tubing, double wall, 119-120, 121
  tubing, single wall, 118-119
  welds, 116-118
  xeroradiography, 156-157
radiographs
  contamination, 55
  defined, 12
  sensitivity of, 87
  typical discontinuities, 42
  usefulness, 41-42
radioisotopes, 26-28, 38-40
  dose rates at 1 ft emissivity, 62*table*

dose rates per curie versus distance, 63*table*
  regulations concerning use and transport of, 38, 57, 68-73
radioscopic digital imaging, 135
radium, as gamma ray source, 26, 38
radium sulfate, 38
Recommended Practice No. SNT-TC-1A, 4-5, 7
regulations, 38, 57, 68-73
rem (roentgen equivalent mammal), 59, 60
resistors, radiography, 132, 133
roentgen (R), 58, 59-60
roentgen equivalent mammal (rem), 59, 60

S

safelights, for darkrooms, 48, 54
safety, 4
  electrical safety, 77
  maximum permissible dose, 60-61
  and Nuclear Regulatory Commission, 57, 68-73
  radiation detection and measurement instruments, 73-77
  radiation dose measurement, 57-60
  radiation protection, 61-68
scanning beam, reversed geometry detectors
  applications, 145-146
  principles, 137, 139-140
scatter, 24
scatter radiation, 15, 24-25
  and contrast, 43
  controlling, 100-101
  and determination of danger zone for gamma radiography, 68
screens, 84-86
security, X-radiography linear detector array application, 151
selenium-75, gamma ray source, 26, 39
  dose rate per curie versus distance, 63*table*
  gamma ray energy, 28
  radioisotope characteristics, 40*table*
semiconductors, radiography, 131-133
sensitivity of radiograph, 87
shallow dose equivalent, 60
shielding, 64-65
shields, 91, 101
shim stock, 88-89
sidescatter radiation, 24-25
sievert (Sv), 58, 60
signals, 72
silver bromide, in radiographic film, 41
  developing, 50-51
  fixing, 52

single surface radiography, of brazed honeycombs, 127, 129

single wall radiography, of tubing, 118-119

SI units, 59-60

slow film, 46, 47, 48

sodium iodide, thallium activated, 139

soft X-rays, 19, 21*table*

source energy, exposure variable, 101

source size, 13-14

    exposure variable, 99

source strength, exposure variable, 101

source-to-film distance, 12

    exposure variable, 99-100

    and sharpness, 13

spatial resolution, in digital radiographic imaging, 141-146

specific activity, 27

spheres, radiography of closed, 120, 121

standard image quality indicators, 86, 87

standing wave, 32

state regulations: agreement states, 38, 57

steel

    choice of filter for, 83

    cobalt-60 for gamma radiography of, 38

    example exposure calculations, 111-115

    gamma ray exposure charts, 95, 96, 107

    iridium-192 for gamma radiography of, 39

    practical thickness limits, 98*table*

    for radiation shielding, 64, 64*table*

    radiographic equivalent factor, 97, 98*table*

    thulium-170 for gamma radiography of, 39

    voltage versus thickness ranges, 108

    X-ray exposure charts, 92, 104

steel filters, 83

stereoradiography, 157-158

stereoscopic radiography, 158

stop bath, 48, 49, 51, 55

stop bath tanks, 55

storage containers, safety requirements, 68, 72

storage phosphors, 139

subject contrast, 43-44

survey meters, 4, 73

    using to ensure safe storage of gamma ray sources, 68

T

tank processing, 49-54

tanks for, 55

tanks, radiography of closed, 122

target, 31

target-to-film distance

    exposure variable, 99

and sharpness, 13, 14

television radiography, 155-156

temperature, of wash tank for developing, 53

temporary job sites, protective shielding issues, 66

tenth value layers, 64*table*, 64-65

thallium activated sodium iodide, 139

thermal neutrons, quality factor value, 58*table*

thermoluminescent dosimeters (TLDs), 71, 73, 75

thin film transistor arrays, 136, 138, 148

360° sweep, 73

thulium-170, gamma ray source, 26, 39

    dose rate at 1 ft emissivity, 62*table*

    dose rate per curie versus distance, 63*table*

    gamma ray energy, 28

    radioisotope characteristics, 40*table*

    voltage versus thickness range for steel, 108

thulium oxide (Tm-203), 39

time, exposure variable, 101

titanium, radiographic equivalent factor, 98*table*

total effective dose equivalent, 60

training, 7, 8

transistors, radiography, 131, 132, 133

travelling wave, 32

tripods, 90

tube current, effect on X-ray intensity and quality, 19, 20, 21*table*

tube envelope, X-ray tubes, 30

tubeheads, for shielding X-ray equipment, 34, 35-36

tubing

    amorphous silicon detector imaging of aluminum, 149

    double wall radiography, 119-120, 121

    single wall radiography, 118-119

tungsten

    as anode material, 31

    for diaphragms, collimators, and cones, 83

U

U channel vertical tie exposure, 130

umbra, 13

United States Coast Guard, enforcement of radioactive material regulations, 57

United States Nuclear Regulatory Commission (NRC), 38, 57, 68-73

United States Nuclear Regulatory Commission Licensing Guide, 68

universal number belts, 90

unsharpness, of images, 13-14

    and source-to-film distance, 99-100

Economy Inn
785 - 448 - 6816

Garnet
785 - 448 - 6800

Continental
Bkfst